OIL AND GAS
PIPELINE
FUNDAMENTALS

OIL AND GAS
PIPELINE
FUNDAMENTALS

JOHN L. KENNEDY

PennWell Books
PennWell Publishing Company
Tulsa, Oklahoma

Copyright © 1984 by
PennWell Publishing Company
1421 South Sheridan Road/P. O. Box 1260
Tulsa, Oklahoma 74101

Library of Congress cataloging in publication data

Kennedy, John L.
Oil and gas pipeline fundamentals

Includes index.
1. Petroleum—Pipe lines. 2. Gas, Natural—Pipe lines.
I. Title.
TN879.5.K35 1984   665.5'44   83-13278
ISBN 0-87814-246-0

Printed in the United States of America

1992

Contents

Preface

A book aiming to describe all phases of oil and gas pipeline design, construction, and operation can only highlight the skills, equipment, and technology required. Pipeline systems in scores of countries around the world differ in purpose, size, complexity, operating environment, regulatory requirements, economic conditions, and design philosophy.

Some aspects of pipeline design and operation are based on physical laws. The relationship between pipeline operating pressure and fluid capacity, for instance, is not affected by political boundaries. Describing such relationships is relatively straightforward.

But how each company chooses to control its pipeline, or regulations governing operation and construction, often can only be introduced by discussing representative situations in a book of this type. For this reason, considerable use is made of examples, rather than attempting to include all possible variations. These examples do not represent the approach of all pipeline builders and operators, but an attempt has been made to choose those that represent accepted technology and equipment.

This book is not a pipeline design manual. Rather, it is written to provide those in other phases of the petroleum industry with a basic knowledge of oil and gas pipeline operations and to familiarize those not involved in day-to-day petroleum operations with the oil and gas pipeline industry.

Despite the introductory purpose of this book, it does contain a handful of equations. These do not by any means provide complete design information, but they are included where appropriate to indicate the many variables in key phases of pipeline design.

Each chapter—indeed, many parts of each chapter—is the focus of large amounts of literature and sizable investments in research and development. Obviously, much detail had to be omitted. But the purpose of this book is to acquaint the nonexpert with oil and gas pipelines and how they tie the world's reserves of oil and gas to the consumer.

A number of references included at the end of each chapter give additional detail on many subjects discussed only briefly here. They can further satisfy an appetite for information about the safe, efficient service provided by the world's vast pipeline network.

Acknowledgments

For their helpful suggestions during the writing of this book, I thank Gene T. Kinney, editor in chief of *Oil & Gas Journal* and formerly pipeline editor of the *Journal,* and Earl Seaton, *Oil & Gas Journal* pipeline editor. Their knowledge of the pipeline industry was a valuable resource.

1

PIPELINE INDUSTRY OVERVIEW

ONE of the most important links in the chain of operations that brings oil and gas from the reservoir to users around the world is a network of pipelines that transports oil, natural gas, and other products from producing fields to consumers. This network gathers oil and gas from hundreds of thousands of individual wells, including those in some of the world's most remote and hostile areas, and eventually distributes a range of products to individuals, residences, businesses, and plants.

This vast gathering and distribution system comprises hundreds of thousands of miles of pipeline—almost a half-million miles in the U.S. alone—varying in size from 2 in. in diameter to as much as 60 in.

Though pumping stations and other facilities are scattered along pipeline routes, most of the world's oil and gas pipeline system is not visible. Pipelines bring oil from Alaska and oil and natural gas from Siberia to consumers. Oil and gas produced from offshore wells are brought to shore by pipeline, often through water several hundred feet deep.

Oil and gas pipeline systems are remarkable for their efficiency and low transportation cost (Fig. 1–1). Just as remarkable is the technology that makes it possible to install large pipelines in areas such as the Arctic permafrost regions and deep water without damage to the environment and with a high degree of safety.

In addition to being a low-cost transportation method, pipelines are energy efficient. A number of studies of energy efficiency have been made and the results vary widely. A recent, more thorough investigation concludes that crude trunk lines consume about 0.4% of the energy content of the crude transported

1

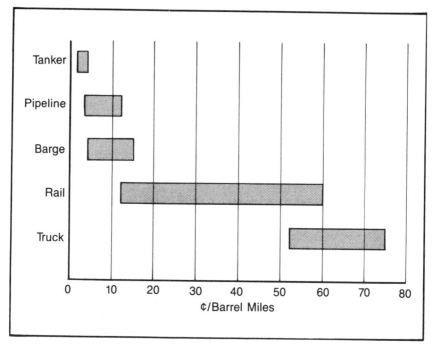

Fig. 1–1.   Petroleum transportation costs.

per 1,000 km (621 miles).[1] Products pipelines use about 0.5% of the energy content of the products moved per 1,000 km. In that report, these rates are compared with estimates of 0.8% for coal trains; 1.0% for oil movement by rail; 2.5% for natural gas pipelines; 3.2% for oil trucks; and 5.4% for coal trucks. Energy consumption for water transport is not known precisely but is estimated to be 0.8% for oil and 1.1% for coal. The study further indicates that crude oil trunk lines consume about 250 BTU/ton-mile; crude oil gathering lines use about 490 BTU/ton-mile; and oil-products pipelines consume about 300 BTU/ton-mile.

According to this report, the amount of energy used depends heavily on the pipeline diameter and on the rate of flow. Energy consumption for crude oil pipelines ranges from about 550 BTU/ton-mile for a 6-in. pipeline to about 180 BTU/ton-mile for a 40-in. diameter pipeline. Energy consumption in products pipelines varies over roughly the same range.

The link between pipeline size and economy is apparent. The relationship between size and capacity is also dramatic. A 36-in. diameter line can carry up to 17 times more oil or gas than a 12-in. diameter pipeline, but construction and operating costs do not increase at nearly the same ratio.

Oil and gas are not the only materials transported by pipeline. Coal and other solids are being successfully pipelined today as well.

History of pipelines

Pipelines to carry water for municipal and home use were built many centuries ago. The history of oil and gas pipelines as they are used today begins shortly after what is considered the first commercial oil well was drilled in Pennsylvania in 1859. However, a natural gas pipeline was reportedly laid in 1825.[2]

The first cross-country oil pipeline was laid in Pennsylvania in 1879, a 109-mile long, 6-in. diameter line from Bradford to Allentown. In 1886, an 87-mile, 8-in. diameter natural gas line from Kane, Pennsylvania, to Buffalo, New York, was built.

In the early 1900s pipeline construction began to expand. In 1906 a 472-mile, 8-in. pipeline from a new field in Indian Territory (Oklahoma) to Port Arthur, Texas, challenged the technology of the time. Early pipelines were built using threaded pipe that workers screwed together with large tongs. It wasn't until about 1920 that welding the separate lengths of pipe together became an accepted construction practice. Oxyacetylene welding was introduced in 1920 but was replaced by the late 1920s with electric welding.[2] Since that time, virtually all oil and gas pipelines of significant diameter and length have been welded. Welding technology has progressed to keep pace with the demands of new pipe steels, increased pipe diameters and pipeline lengths, and the challenges of offshore and Arctic environments.

Some significant early pipeline projects included a pipeline 250 miles long built from the Texas Panhandle to Wichita, Kansas, in 1927, later extended to Kansas City, Missouri. And in 1928, a joint-venture company built a crude oil pipeline from Oklahoma to Chicago.

Cross-country pipelines to carry products got their start in 1930 when a group of midcontinent U.S. refiners built a pipeline network to deliver refined products that allowed them to compete in markets such as Chicago, Minneapolis and other cities.

Long-distance oil and gas pipeline transportation got a boost during World War II when coastal tanker traffic was disrupted. A way to move oil and products from fields in the Southwest United States to the East Coast was needed. Two pipelines were built: one was a 1,250-mile, 24-in. diameter crude oil line, the other a 1,470-mile, 20-in. diameter products line. At about the same time these government-owned lines were built, Tennessee Gas Co. built a 24-in. diameter, 1,265-mile natural gas line from the Southwest to the East Coast. In 1947, Texas Eastern bought the government lines and converted them to natural gas, connecting the existing electric motors to centrifugal compressors for the first time in a long-distance pipeline. Also that year, the first line from the Southwest United States to California was built.

In the early 1950s, major pipelines were built in Canada and the Trans-Arabian Pipe Line was constructed from the Persian Gulf to the Mediterranean Sea.

In the 1960s, larger-diameter pipelines proved their economic advantage. A products line consisting of 32-in., 34-in., and 36-in. pipe was built from Houston to New York to break the bottleneck created by striking maritime unions, and a 40-in. crude pipeline was constructed from Louisiana to Illinois. Operating costs on Colonial's Houston-New York products line were reported to be 11¢/1,000 bbl-miles, compared with 37.3¢/1,000 bbl-miles for the next largest pipeline.[2] Then, Colonial was the largest privately financed project ever built; later the trans-Alaska line set a new record.

The world's pipeline network expanded rapidly when it became apparent that pipelines were an efficient, economic way to move oil, gas, and products to consumers. Contributing to the need for expansion were large new discoveries of oil and gas, many of them in remote areas that had little local demand. Pipelines were needed to move those supplies to markets.

In the North Sea, for example, a vast offshore complex of oil and gas fields developed during the 1960s and 1970s. The area was a proving ground for much of today's offshore pipeline construction technology. Discovery of oil on Alaska's North Slope in the late 1960s called for a pipeline that had to be built and operated under conditions never before encountered.

Discoveries of oil and gas will continue to be made in remote areas and in hostile environments. On land and on the world's continental shelves, the preferred method for moving those supplies to market will be a pipeline. Tankers are necessary for long-distance ocean transportation, but they are only a link in the gathering and distribution chain. Pipelines must gather oil and gas to the tanker-loading port and distribute the cargo after the tanker has reached its destination.

Though it apparently is only coincidence, many of the world's large supplies of oil and gas are far from populated areas. In the United States, Texas, Louisiana, and Oklahoma are major oil- and gas-producing states. Production there is much more than enough to supply the needs of those states, and the surplus must be moved to large consuming areas. An even more dramatic example is Alaska's North Slope where the largest oil field in the United States is located in one of the most sparsely populated states. The innovative, costly trans-Alaska pipeline was required to move that oil to a tanker port for transportation to large markets.

There are many other examples of this situation worldwide. Huge reserves of oil and gas exist in the Middle East and in remote areas of the Soviet Union. Making use of that energy requires a complex system of pipelines and tanker routes.

Even more challenging projects face the industry in the 1980s. If natural gas from the Arctic is brought to consumers in Canada and the United States, a pipeline network several thousand miles long will be needed. The cost, according to one estimate, would be between $25 and $30 billion. Such pipelines pose many technological obstacles, but economics is often the biggest hurdle. When the capital cost of a pipeline is so high, the transportation cost that must be added to the cost of the gas can put the delivered cost out of the reach of many customers. Especially if gas continues to be found in more accessible areas, such high-cost gas will not be competitive in the energy market.

The future of oil and gas pipelines will be marked by additional technological and economic challenges. Pipelines will likely be laid in greater water depths, and they will be used increasingly to carry other materials—coal, for instance. The steadily rising cost of fuel, materials, and labor will also pressure builders and operators of future pipelines to improve efficiency. Rising costs, magnified by increased activity in hostile environments, may make cost cutting the biggest challenge in the pipeline industry's future.

Supplies and markets

For a perspective on worldwide pipeline transportation of oil and gas, it is necessary to know where significant oil and gas supplies exist, relative to markets.

Table 1–1 shows a breakdown of the major producing countries and the major consuming areas. It is apparent that in many cases, both tanker and pipeline transportation are required. The Middle East, the largest producing area outside the Soviet Union, ships much of its crude oil by tanker to Europe and the United States. But a huge network of pipelines is necessary to move crude from producing wells to the tanker port for shipment. Other pipelines are needed in the consuming countries to move crude to refineries for processing.

One of the largest natural gas supplies is in the Soviet Union's western Siberia area. In 1982 construction was underway on a large-diameter pipeline system to move gas from that area, including a large-diameter pipeline almost 2,900 miles long to export gas to Western Europe.

Work has been underway for some time on a pipeline to transport natural gas from Algeria to Italy and other areas of Europe. Portions of that system cross the Mediterranean and the Messina Strait in water depths up to 1,968 ft.

The trans-Alaska pipeline, now moving oil 900 miles from Alaska's North Slope to a tanker terminal at Valdez, was built at a cost of about $10 billion. It had to meet environmental restrictions never before imposed on such a project and was constructed under some of the most difficult conditions ever faced by pipeline builders.

These examples are by no means the only significant pipelines in existence or planned, but they indicate how frequently significant oil and gas supplies are found in areas remote from markets. They also indicate the magnitude of many pipeline projects and the huge costs that are often involved.

Just as important are less dramatic supply routes, such as the many pipelines from the southwestern and Gulf Coast areas of the United States to large

TABLE 1-1
Outlook For Non-Communist Oil

	1979 average	1980 average	1981 average	1982 average	1983 average
			Million b/d		
CONSUMPTION					
United States*	18.5	17.1	16.0	15.3	15.4
United States territories	0.4	0.4	0.3	0.3	0.3
Western Europe	14.8	13.5	12.5	11.9	12.0
Japan	5.5	5.0	4.7	4.5	4.5
Other IEA†	2.7	2.6	2.4	2.4	2.4
Subtotal IEA	39.1	35.8	33.6	32.4	32.6
Subtotal OECD	41.8	38.5	35.9	34.4	34.6
Other developed countries‡	0.4	0.4	0.4	0.4	0.4
OPEC	2.5	2.7	2.8	2.9	3.1
Non-OPEC LDCs	7.5	7.7	7.8	7.9	8.1
Offshore military and other	0.2	0.2	0.2	0.2	0.2
Total	52.4	49.5	47.1	45.8	46.4
Inventory change	+0.7	+0.6	-0.7	-2.2	+0.5
Total supply	53.5	50.1	46.4	43.6	46.9
End of period inventory (million bbl)	5,150	5,354	5,120	4,345	4,530
SUPPLY					
Non-OPEC§					
United States‖	10.7	10.8	10.7	10.6	10.5
Canada	1.8	1.8	1.6	1.5	1.6
North Sea	2.1	2.2	2.4	2.6	2.7
Other developed countries	0.9	0.8	0.9	0.9	0.9
Mexico	1.6	2.1	2.6	2.9	3.2
Other LDCs	3.6	3.6	3.7	3.9	4.0
Subtotal	20.7	21.3	21.8	22.4	22.9
OPEC					
Algeria	1.2	1.0	0.8	0.7	—
Ecuador	0.2	0.2	0.2	0.2	—
Gabon	0.2	0.2	0.2	0.2	—
Indonesia	1.6	1.6	1.6	1.5	—
Iran	3.2	1.5	1.3	1.5	—
Iraq	3.5	2.6	1.0	1.0	—
Kuwait	2.2	1.4	0.9	0.8	—
Libya	2.1	1.8	1.1	0.9	—
Neutral Zone	0.6	0.5	0.4	0.3	—
Nigeria	2.3	2.1	1.4	1.5	—
Qatar	0.5	0.5	0.4	0.4	—
Saudi Arabia	9.2	9.6	9.6	7.4	—
United Arab Emirates	1.8	1.7	1.5	1.2	—
Venezuela	2.4	2.2	2.1	1.7	—
NGL	0.8	0.9	0.9	1.0	—
Subtotal	31.7	27.8	23.6	20.3	23.3
Net Communist exports	1.1	1.0	1.0	0.9	0.7
Total supply	53.5	50.1	46.4	43.6	46.9

*Includes processing gain. †Canada, Australia, and New Zealand. ‡Israel and South Africa. §Including natural gas liquids. ‖Including processing gain.

Source: Department of Energy and *Oil & Gas Journal,* 3 May 1982, p. 96

consuming and refining areas in the Midwest and the East. For years these pipelines have made needed supplies available at costs that could not be equaled by any other means of transportation.

The parties involved

Oil, natural gas, and natural gas liquids pipeline gathering and transportation systems are owned by several types of firms. In the United States, where complex regulations exist governing the transportation and pricing of energy, most gathering and long-distance pipelines are owned by pipeline companies whose sole function is to operate a pipeline system. Though these companies may have names similar to, or may be a subsidiary of, an oil- or gas-producing company, the pipeline company is normally a separate legal entity. Outside the United States, a wide variety of relationships exist in the ownership of pipelines, depending in part on the individual country's laws.

In the United States, natural gas is usually purchased by the pipeline company and is resold to industry and distribution companies. Oil is transported via pipeline by a shipper-owner, usually a refiner.

Pipeline construction around the world is usually done by a construction contractor rather than by the pipeline owner. The need for costly, highly specialized equipment and specialized talent makes this the best approach. Pipeline owners could not afford to own and maintain the required equipment and use a construction staff for only intermittent construction work on their own systems. The pipeline contractor can utilize his equipment and staff more efficiently by doing work for many pipeline owners.

In addition to the construction contractors who install pipelines both on land and offshore, there is a long list of firms providing specialized services such as pipe coating, maintenance, facilities design, and inspection. An even longer list of manufacturers and suppliers provides equipment, including valves, compressors, pumps, instruments, prime movers, controls, and maintenance tools.

Key industry statistics

Exact details of the world's total pipeline lengths, sizes, and capacities are not available. In the United States, data are maintained on all interstate pipelines and related facilities regulated by the federal government. These data offer insight into the efficiency of a vast network of oil and gas pipelines.

In 1981, U.S. interstate pipelines delivered more than 10 billion bbl of crude and products and 18.5 trillion cu ft of gas.[3] The investment in equipment related to these liquids and gas pipelines was estimated at $60 billion, and combined United States pipeline mileage stood at 447,449 miles (Table 1–2). Liquids pipeline companies operated more than 38,000 miles of gathering lines, almost 58,000 miles of crude oil trunk lines, and over 76,000 miles of petroleum

TABLE 1–2
United States Interstate Pipeline Mileage: 1972–1981

Year	Gas	Miles Liquids	Total
1972	246,551	173,532	420,083
1973	248,773	170,691	419,464
1974	246,872	173,342	420,213
1975	255,792	172,680	428,472
1976	254,548	174,072	428,590
1977	256,198	170,948	427,146
1978	258,576	162,452	421,028
1979	264,134	169,794	433,928
1980	274,248	172,673	446,921
1981	274,634	172,815	447,449

Source: FERC Form P and Form 2 annual reports filed by interstate pipelines and FERC published reports, *Oil & Gas Journal,* 22 November 1982, p. 73

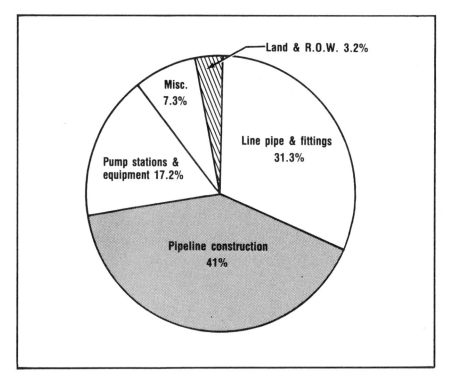

Fig. 1–2.   Investment in crude oil pipelines. Source: *Oil & Gas Journal,* 22 November 1982, p. 73.

products lines. Interstate natural gas pipeline companies operated about 197,000 miles of transmission pipelines, 73,000 miles of field gathering lines, and 4,600 miles of pipeline associated with storage facilities during 1981.

Of the approximately $60 billion in property operated by all interstate pipelines, crude and products pipelines accounted for $21.2 billion.[3] The share of this total investment represented by crude oil pipelines included 31.3% for line pipe and fittings, 41% for pipeline construction, 17.2% for pump stations and equipment, 3.2% for land and right of way, and 7.3% for miscellaneous costs (Fig. 1–2). The investment split for interstate products pipelines was 30.4% for line pipe and fittings, 40% for pipeline construction, 15.5% for pump stations and equipment, 3% for land and right of way, and 11% for miscellaneous (Fig. 1–3).

In 1981, 132 interstate liquids pipeline companies filed reports with the Federal Energy Regulatory Commission (FERC); 104 interstate natural gas companies reported. Operating revenue of interstate crude and products pipe-

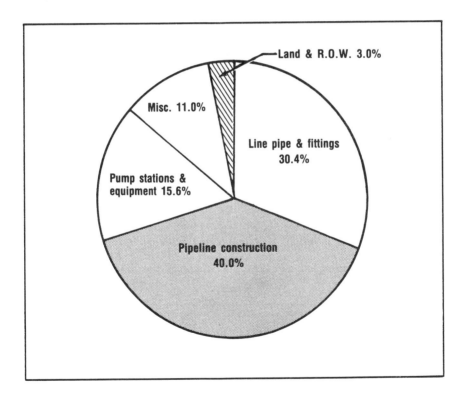

Fig. 1–3.   Investment in products pipelines. Source: *Oil & Gas Journal*, 22 November 1982, p. 73.

lines was about $6.7 billion; operating revenue of natural gas pipelines was about $58 billion.

Interstate pipeline companies in the United States are only part of worldwide oil and gas pipeline operations. The industry also includes pipeline owners in all oil and gas-producing countries, pipeline construction firms that own and operate equipment valued at billions of dollars, and manufacturers and service companies.

Yearly pipeline construction. Tens of thousands of miles of pipelines are built each year around the world (Fig. 1–4). The amount of pipeline construction depends in part on the number of oil and gas discoveries and their location. Construction activity is also related to the need to obtain permits, political obstacles, and the availability of equipment and supplies. For example, construction of the trans-Alaska crude pipeline did not begin for more than nine years after oil was discovered on the North Slope because objections to the project on environmental grounds delayed government approval. Other projects are often delayed because of financing problems.

But these unique projects represent only a part of the pipeline construction work done each year. Many thousands of miles of smaller diameter pipeline are laid all over the world to move oil and gas from producing wells to refineries and

Fig. 1–4. Sideboom tractors prepare to move pipe for marsh crossing. Source: *Oil & Gas Journal*, 20 November 1978, p. 110.

processing facilities. Many of these individual pipelines are only a few miles long, or less.

In 1983, it was estimated that the industry would lay more than 27,000 miles of pipeline worldwide in that year alone, including oil- and gas-gathering and transmission lines, and products pipelines.[4] This total did not include pipelines to be laid in Communist countries. Also, these estimates did not include construction of distribution lines to deliver natural gas to individual homes and businesses.

Laying the estimated 27,000 miles of pipeline in 1983 was expected to cost about $18.5 billion. Of this total, new onshore pipelines were expected to cost about $14 billion, including sizes ranging from 4-in. diameter to more than 30-in. diameter. Offshore pipeline construction during 1983 was to cost about $4.5 billion. Table 1–3 shows details of the 1983 estimated pipeline construction.

Most oil and gas pipelines are built in producing countries, and a large share of all oil- and gas-gathering and transmission lines is built in the United States. Of the more than 18,000 miles of gas pipeline scheduled for construction in 1983, for example, about 7,800 miles were to be built in the United States. About 3,000 miles of that total were to be 4-in. to 10-in. diameter. More than 2,600 miles were planned in sizes between 22 in. and 30 in.

Pipeline construction activity also is at a high level when a country is in the early development of large oil and gas reserves. For example, in the late 1970s and early 1980s, large reserves were discovered in Mexico that required a network of pipelines to connect producing wells with refineries and processing plants.

In the North Sea, a large amount of pipeline construction was related to development of huge fields. Many of those pipelines were long and had to be laid in deep water under difficult conditions. Other areas where pipeline construction is important include the Middle East, Africa, Canada, and the Asia/Pacific area. Major projects can cause variations in year-to-year activity in any area, but wherever oil and gas reserves exist and aggressive exploration and development programs are carried out, oil and gas pipeline construction is an important activity.

Example pipeline costs. There is no typical pipeline as far as cost is concerned. Construction costs depend on geographical area, size of pipeline, location on or offshore, number and size of pump stations or compressor stations and related facilities, and general economic conditions.

In general, the longer the pipeline, the lower the cost per mile. A pipeline a few miles long usually costs considerably more per mile than a pipeline several hundred miles long, even if both are the same diameter and are laid in a similar environment. Of course, total cost of the longer pipeline will be greater, other factors being equal.

TABLE 1-3
Non-Communist Pipeline Construction In 1983

	4–10 in.	12–20 in.	22–30 in.	Larger than 30 in.	Total
			Miles		
GAS PIPELINES					
United States	2,965	1,085	2,610	1,110	7,770
Canada	820	134	188	829	1,971
Latin America	131	764	1,650	268	2,813
Asia-Pacific	242	236	2,563	79	3,120
Western Europe	75	295	1,068	802	2,240
Middle East	68	26	—	205	299
Africa	105	55	—	—	160
Total gas	4,406	2,595	8,079	3,293	18,373
CRUDE PIPELINES					
United States	288	419	—	798	1,505
Canada	—	542	—	—	542
Latin America	224	366	—	—	590
Asia-Pacific	5	—	—	—	5
Western Europe	4	351	—	—	355
Middle East	276	—	307	28	611
Africa	—	—	1,220	—	1,220
Total crude	797	1,678	1,527	826	4,828
PRODUCT PIPELINES					
United States	506	291	623	391	1,811
Canada	94	—	—	—	94
Latin America	435	1,568	—	—	2,003
Asia-Pacific	31	15	—	—	46
Western Europe	—	—	—	—	—
Middle East	—	29	—	—	29
Africa	113	37	317	—	467
Total product	1,179	1,940	940	391	4,450
NON-COMMUNIST WORLD TOTALS					
Gas	4,406	2,595	8,079	3,293	18,373
Crude	797	1,678	1,527	826	4,828
Product	1,179	1,940	940	391	4,450
Total	6,383	6,213	10,546	4,510	27,651

Source: *Oil & Gas Journal*, 24 January 1983, p. 25

Pipeline costs are sometimes compared on an "inch-mile" basis to make the comparison less dependent on pipeline size. Compressor costs are often stated in $/horsepower to permit comparing compressors of different size. As in the case of a pipeline, a large pump or compressor station costs less per horsepower than a smaller station, though total cost is greater.

Generally, costs for an offshore pipeline are much higher than for a pipeline on land. But in extreme environments, such as the Arctic or mountainous regions, land pipeline construction costs can also be very high.

Construction applications filed with FERC in the United States contain cost data that provide some representative pipeline building costs. These data indicate that for onshore pipelines (Fig. 1–5), material accounts for 44.6% of

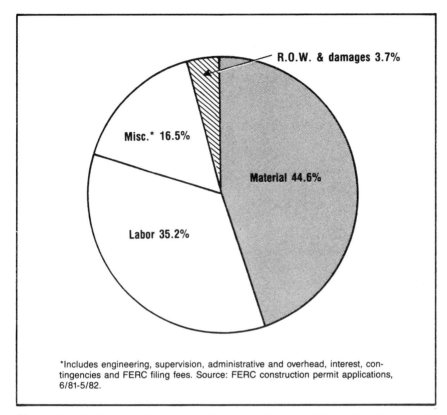

R.O.W. & damages 3.7%

Misc.* 16.5%

Material 44.6%

Labor 35.2%

*Includes engineering, supervision, administrative and overhead, interest, contingencies and FERC filing fees. Source: FERC construction permit applications, 6/81-5/82.

Fig. 1–5. Onshore pipeline construction cost. Source: *Oil & Gas Journal*, 22 November 1982, p. 73.

total construction cost; labor, 35.2%; right of way and damages, 3.7%; and miscellaneous costs, 16.5%. Miscellaneous expenses include engineering, supervision, administration and overhead, interest, contingencies, and filing fees. Offshore construction costs (Fig. 1–6) from FERC construction applications include 26.5% for material, 57.2% for labor, and 16.3% for miscellaneous expenses.

Completed cost data reported to the FERC in 1981 indicate average United States pipeline construction cost ranged from about $186,000/mile for an 8-in. diameter pipeline to almost $900,000/mile for a 36-in. diameter pipeline.[3] These data are averages, and the range of costs for individual projects within each pipeline size varies greatly.

Onshore compressor stations completed in the United States and reported to FERC ranged in cost from just over $500/horsepower to almost $2,000/installed horsepower. These costs represent new stations in which reciprocating compres-

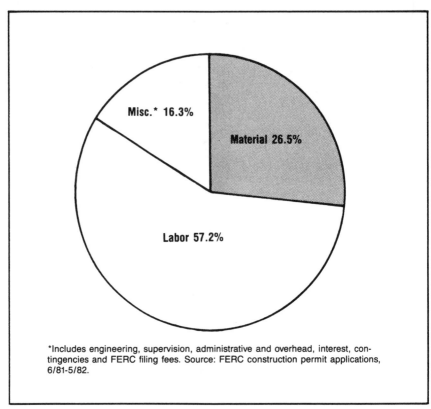

*Includes engineering, supervision, administrative and overhead, interest, contingencies and FERC filing fees. Source: FERC construction permit applications, 6/81-5/82.

Fig. 1–6. Offshore pipeline construction cost. Source: *Oil & Gas Journal,* 22 November 1982, p. 73.

sor units were installed. Where additional compressors were installed at existing stations, the $/horsepower cost is generally lower because much of the auxiliary equipment is already installed. The cost of additions to onshore compressor stations reported to the FERC ranged from $498/installed horsepower to $1,361/ installed horsepower.

A breakdown of onshore compressor station cost shows that equipment and materials account for 56.9% of the total; installation labor, 20.3%; and land, 1.2%. Miscellaneous costs, including engineering, supervision, administration and overhead, interest, contingencies, and filing fees, account for 21.6% of the total. Offshore compressor station costs, according to FERC reports, include 60% for equipment and materials, 24.8% for installation labor, and 15.2% for miscellaneous charges. Individual project costs vary widely, but some examples of project costs are shown in Tables 1–4 through 1–7.

To help project pipeline construction costs, indices for both oil and natural gas pipelines have been developed. The OGJ-Morgan indices for oil pipelines

TABLE 1–4
520 Miles of 24-in., Wyoming, Colorado, and Kansas

Item	Material	Installation	Total
Right of way and damages		$ 6,340,000	$ 6,340,000
Survey and mapping		1,820,000	1,820,000
520 miles of 24-in. OD, 0.257-in. W.T.			
X-65 pipe	$68,640,000	54,912,000	123,552,000
Coat and wrap	4,530,000		4,530,000
Cathodic protection	48,500	62,400	111,000
Road and railroad crossings	124,200	120,600	245,000
Concrete pipe weights—812 set-on	178,640	162,400	341,000
Concrete pipe weights—184 bolt-on	58,880	64,400	123,000
X-ray		3,536,000	3,536,000
River crossing—1		460,000	460,000
Thirty-two mainline valves	1,856,000	1,280,000	3,136,000
Four scraper trap assemblies	1,916,000	916,000	2,832,000
Tie-in	167,000	155,000	322,000
Engineering, supervision and inspection		11,700,000	11,700,000
Subtotal	77,519,220	81,528,000	159,048,000
Omissions and contingencies			7,952,400
Total direct cost			167,000,400
Administrative overheads			4,175,010
Allowance for funds used during			
construction			12,153,450
Total cost			$183,328,860
Cost/mile			$352,556

Source: *Oil & Gas Journal,* 22 November 1982, p. 73

and for gas pipelines are similar, but differences in some cost elements make separate indices desirable. Operating pressures are higher in natural gas pipelines, for instance, influencing the cost of pipe and other equipment. In addition, safety codes and requirements are more stringent for natural gas pipelines. Also, a separate cost index is needed because items such as compressors, dehydrators, separators, and scrubbers are not used in oil pipeline systems.

These indices do not give dollar values for individual components. Rather, they are designed to show cost trends to aid in projecting future construction costs and to help estimate the value of existing facilities. These indices are computed for the major components of construction (Table 1–8).[5]

Operating and maintenance expenses are substantial for both natural gas and liquids pipelines. For instance, statistics show that the operating expense for interstate natural gas pipelines totaled about $2.253 billion, or more than $11,711/mile of installed pipeline in 1980.[6] Maintenance expense for interstate natural gas pipelines was about $1,526/mile, for a total of almost $300 million (Table 1–9).

TABLE 1-5
21.1 Miles of 12-in., Oklahoma

Item	Cost
Survey	$ 103,000
Right of way and damages	378,000
Subtotal	481,000
MATERIAL	
19.5 miles 12-in. OD, 0.281-in. W.T. X-52 pipe	1,445,000
1.6 miles 12-in. OD, 0.325-in. W.T. Grade B pipe	167,000
Coating	228,000
Valves	138,000
Other material	869,000
Subtotal	2,847,000
INSTALLATION	
Direct	2,105,000
Indirect	712,000
Subtotal	2,817,000
Contingency	430,000
General plant	131,440
Allowance for funds used during construction	295,870
Cost of certificate	6,640
FERC filing fee	13,640
Total cost	$7,022,590
Cost/mile	$332,824

Source: *Oil & Gas Journal,* 22 November 1982, p. 73

TABLE 1-6
28.28 Miles of 20-in., Offshore Louisiana

Item	Cost
Right of way	$4,170
MATERIAL	
27.53 miles 20-in. OD, 0.438-in. W.T. pipe	6,540,300
0.18 miles 20-in. OD, 0.500-in. W.T. pipe	52,800
0.57 miles 20-in. OD, 0.562-in. W.T. pipe	195,000
Platform piping	260,700
Coating, riser, fittings, valves, etc.	1,599,360
Subtotal	8,648,160
INSTALLATION	
Install 20-in. OD line	5,874,060
Install 20-in. riser	292,900
Install platform piping	341,000
Install tie-in assemblies	1,000,000
Platform space rights	330,000
Diving, helicopter, boat services, coating pipe, etc.	669,600
Subtotal	8,507,560
Surveys	787,160
Field engineering and supervision	724,780
Franchise and consents	65,950
Overhead	945,510
Allowance for funds used during construction	459,519
Contingencies	1,015,181
Total cost	$21,157,990
Cost/mile	$748,161

Source: *Oil & Gas Journal,* 22 November 1982, p. 73

TABLE 1-7
Construct 38,000-hp Compressor Station Onshore

Item	Material	Installation	Total
Land	$ 240,161		$ 240,161
Structures	1,051,905	270,181	1,322,086
Yard improvements	57,639	121,281	178,920
Station piping	7,334,518	4,931,707	12,266,225
Main compressor units	15,490,387	360,242	15,850,629
Electrical equipment	336,225	186,125	522,350
Other equipment	4,779,205	922,218	5,701,423
General plant	50,434		50,434
Concrete	446,700	1,044,701	1,491,401
Contractor's overhead and fee	8,406	36,024	44,430
Field engineering and supervision		1,555,043	1,555,043
Subtotal	$29,795,580	$9,427,522	$39,223,102
Franchise and consents			91,686
Overhead			2,211,948
Allowance for funds used during construction			5,310,031
Contingencies			2,341,767
FERC fee			2,262
Total cost			$49,180,798
Cost/hp			$1,294

Source: *Oil & Gas Journal*, 22 November 1982, p. 73

TABLE 1-8
OGJ Morgan Pipeline Cost Indexes (1974 = 100)

Cost component	Yearly indexes				1982 quarterly index			
	1978	1979	1980	1981	1st	2nd	3rd	4th
Steel line pipe	160	172	197	219	219	219	208	...
Line pipe fittings	159	169	193	214	216	216	207	...
Pipeline construction	140	156	174	195	197	201	206	...
Pipe coating	135	148	163	181	183	186	190	...
Masonry buildings	127	138	149	162	164	166	169	...
Metal or asbestos clad steel frame buildings	128	137	147	159	161	163	166	...
Stationary engines and reciprocating pumps	156	176	197	220	225	228	232	...
Electric motors 100 hp and up and large centrifugal pumps	185	213	234	256	263	266	269	...
Portable and miscellaneous units	156	175	194	213	217	220	222	...
Electric starters and switchgear	134	143	156	172	176	177	178	...
Shop machinery and tools	155	170	192	215	218	220	223	...
Oil pipe and fittings in place	149	163	184	205	207	209	207	...
Steel storage tanks	157	174	190	208	210	212	215	...
Firewalls and miscellaneous	157	174	190	208	210	212	215	...
Radio and microwave equipment	121	127	143	157	159	161	162	...
Office furniture and equipment	130	141	150	162	164	166	167	...
Autos, tractors, and trucks	135	151	168	183	185	188	190	...
Other work equipment	155	171	193	213	215	217	220	...
Total composite index	150	164	185	205	206	208	206	...

Source: *Oil & Gas Journal*, 9 May 1983, p. 119

TABLE 1-9
Transmission Expense, Interstate Gas Pipelines—1980

	Expense, $	Cost/mile, $	Cost/MMcf sold, $
OPERATION EXPENSES			
Supervision and engineering	100,288,390	521.2	5.78
System control and load dispatching	19,258,089	100.1	1.11
Communication system expense	15,983,714	83.1	0.92
Compressor station labor and expenses	194,241,793	1,009.5	11.20
Gas for compressor station fuel	816,329,988	4,242.5	47.08
Other fuel and power for compressor stations	37,009,308	192.3	2.13
Mains	146,269,485	760.2	8.44
Measuring and regulating station expenses	53,934,064	280.3	3.11
Transmission and compression of gas by others	825,545,590	4,290.4	47.61
Other transmission expenses	31,074,869	161.5	1.79
Rents	13,544,364	70.4	0.78
Total operation expenses	2,253,479,654	11,711.4	129.96
MAINTENANCE EXPENSES			
Supervision and engineering	24,498,827	127.3	1.41
Structures and improvements	21,090,012	109.6	1.22
Mains	68,896,581	358.1	3.97
Compressor station equipment	154,967,390	805.4	8.94
Measuring and regulating station equipment	11,036,354	57.4	0.64
Communication equipment	9,241,663	48.0	0.53
Other equipment	3,963,906	20.6	0.23
Total maintenance expenses	293,694,732	1,526.3	16.94
Total transmission expenses	2,547,174,386	13,237.7	146.90
Total miles of transmission pipeline	192,418		
Total natural gas sold, MMcf	17,340,086		

Source: Statistics of Interstate Natural Gas Pipeline Companies—1980, United States Department of Energy, *Oil & Gas Journal*, 22 November 1982, p. 73

United States regulation

Pipeline operations in the United States differ from most other countries. Long-distance transportation of natural gas by interstate pipelines and the distribution of natural gas to consumers is carried out mainly by private companies and is regulated under the principles of public utility law. Interstate pipelines are regulated by the Federal Energy Regulatory Commission; distribution companies are regulated primarily by their state public utilities commissions. These agencies determine the price that transmission and distribution companies can charge buyers for natural gas, govern the financial structure of the companies, including permissible profit ranges, and regulate other aspects of operation.

The wellhead price of gas, the price at which producers sell to natural gas pipeline companies, is also currently regulated by the FERC. Wellhead price regulation began in 1954 as a result of a U.S. Supreme Court decision. Price regulation is considered to have had a negative influence on the search for new gas supplies because the price was held below that needed to make drilling and development profitable. In 1978, Congress passed the Natural Gas Policy Act (NGPA) in an effort to provide more incentive to producers to search for new reserves. The act created several categories of natural gas. Some were still to be regulated, some were to be deregulated in 1985, and some categories were immediately deregulated. In the early 1980s, there was considerable debate concerning whether to deregulate all natural gas in the United States immediately or to continue the schedule set forth in the NGPA.

A provision of many gas pipeline contracts called "take or pay" was also the subject of discussion. Historically, when a producer discovered natural gas, he sold it under a long-term contract to a pipeline company. The contract required the pipeline company to purchase the gas at a specified rate, or "take." Even if the pipeline company did not accept delivery of natural gas from the producer, the pipeline company had to pay the producer. The take-or-pay provision was insisted upon by the producer because it ensured a constant market for the gas. He assumed this constant market when agreeing on a firm price over the life of the contract.

Under most such contracts, the pipeline company could recover the gas paid for, but not taken, by taking more than the contract volume over a specified period. As a result of changes in the natural gas market in the 1970s, pipeline companies found themselves with a surplus of gas but were still required to pay for high rates not being taken. Efforts to change laws to reduce the effect of these high take-or-pay rates were underway in 1983.

Most long-distance crude oil pipeline companies in the United States only provide a transportation service; they do not sell energy as gas transmission and distribution companies do. Because of this, most oil pipelines are common carriers. For many years, they were regulated by the Interstate Commerce Commission, but oil pipelines are now regulated by the FERC. In this capacity, the FERC reviews and issues orders establishing tariff rates, investigates newly filed rates for legality, and can order reparations for damages sustained by shippers in the pipeline due to violations of the Interstate Commerce Act.

Under this act, a common carrier pipeline's tariffs and other rules must be reasonable, must apply to all shippers on the line on a nondiscriminatory basis, and must be filed with FERC before petroleum can be shipped under the tariff. A common carrier pipeline is prohibited from giving unreasonable preference or discriminating in any way in furnishing services to different shippers.

Pipelines that carry oil or gas within a single state must comply with that state's regulations. More flexibility in pricing and other financial matters is possible in the case of these intrastate pipelines, but regulations governing construction and operation are usually no less strict than those for interstate pipelines.

There are many other regulations, both state and federal, with which pipeline companies must comply, including environmental restrictions on emissions into the air and discharges into streams; acceptable construction methods for crossing streams, certain types of soils, and specially designated areas; and design criteria for pipelines passing through populated areas.

Safety regulations. In the U.S. the safety of natural gas and liquids pipelines is the responsibility of the Department of Transportation (DOT). The Materials Transportation Bureau of DOT is charged with enforcing pipeline safety under the Natural Gas Pipeline Safety Act of 1968 as amended, the Hazardous Liquid Pipeline Safety Act of 1979, and the Hazardous Materials Transportation Act as amended. Hazardous liquids are defined in the regulations as petroleum, petroleum products, and anhydrous ammonia.

Rules for the design, construction, and operation of natural gas and liquids pipelines to ensure they are safe are set forth in Subchapter D of the Code of Federal Regulations, Title 49—Transportation. Part 192 contains minimum federal safety standards for the transportation of natural gas and other gas by pipeline; Part 195 covers safety standards for the transportation of hazardous liquids.

Regulations regarding pipe steel, welding procedures, installation procedures, compressor shutdown systems, and many more areas are contained in Parts 192 and 195. Part 190 describes procedures for reporting leaks, making inspections, warnings of violations and how to respond, compliance orders, and civil and criminal penalties that can be assessed for violation of CFR 49 regulations. It also describes the authority of the Office of Operations and Enforcement of the Materials Transportation Bureau and how that authority is exercised. Part 191 of CFR 49 explains procedures for reporting leaks in natural gas pipelines.

Regulations in CFR 49 are detailed and specific. Key subjects covered in regulations in Part 192, Minimum Federal Safety Standards for Natural Gas Pipelines, include the following:

1. Materials—including steel pipe, cast iron pipe, plastic pipe, and copper pipe; marking of materials; and pipe transportation.
2. Pipe design—design formulas, yield strength, wall thickness, and design factors for steel pipe; and the design of pipe made from other materials.
3. Design of pipeline components—valves, flanges, other fittings, welded components, compressor stations, instrumentation and control equipment, pressure relief and pressure limiting devices, and others.
4. Welding of steel in pipelines—qualification of welding procedures and welders, preparation for welding, preheating, stress relieving, inspection, testing, and repair.
5. Joining of materials other than by welding—cast iron, plastic, and other types of materials, including procedures and inspection.

6. Requirements for corrosion control—external corrosion control methods, monitoring, inspection; internal corrosion control; and atmospheric corrosion control.

Also covered in this part of CFR 49 are regulations covering customer meters, service regulators, and service lines; test requirements; uprating; operations; and maintenance of natural gas pipelines.

Part 195, covering the transportation of hazardous liquids by pipeline, includes sections on accident reporting, design requirements, construction, hydrostatic testing, and operation and maintenance. Many areas discussed in the liquids pipeline safety regulations are similar to those in Part 192 for natural gas pipelines.

Tanker transportation

Not all oil and gas is moved by pipeline. Many producing countries export large quantities of crude. Ocean transport is necessary to deliver crude and products to refiners and other customers in consuming countries. Crude, products, and natural gas are all moved by tanker, but shipments of crude are by far the largest.

Natural gas must be liquefied when moved by tanker in order to carry a large enough volume to be practical. Tanker movement of LNG requires a liquefaction facility in the exporting country and a gasification facility in the receiving country. Both facilities are typically located near the tanker loading and unloading ports.

The first shipments of sizable volumes of refined petroleum products were made from the United States to England in the early 1860s. Since the 1950s, the size of crude tankers has increased steadily. The largest tankers have gone from 30,000 deadweight tons (dwt) in 1950 to over 500,000 dwt in 1976.[7] Deadweight ton (dwt), a term used to rate tanker capacity, indicates the amount of cargo that can be carried. It is the displacement of the tanker when loaded minus the displacement "light," in tons of 2,240 lb. A deadweight ton is equal to about 7–7½ bbl of crude, depending on the specific gravity of the crude.

Traditionally, much of this tanker transportation system has been owned and operated by oil companies. But in recent years, some oil-producing countries have developed their own national shipping fleet and operated their own tankers. This trend, however, has been limited by the tanker surplus and low rates.

Crude oil. Crude tankers are involved in a significant share of the world's oil movements. In the United States in 1980, pipelines handled about 46% of the crude and products moved, while water carriers accounted for about 25% of the traffic (Table 1–10). The remainder of the crude and products shipped were handled by motor carriers and railroads. Shipments of oil from Alaska have increased the share of crude handled by tankers in recent years.[8] Because most

TABLE 1–10
Breakout of Crude and Products Carried in United States Transportation Systems

Year	Total carried (tons)	— Pipelines — Tons carried	Percent of total	— Water carriers — Tons carried	Percent of total	— Motor carriers* — Tons carried	Percent of total	— Railroads — Tons carried	Percent of total
1975	1,831,515,800	879,449,300	48.02	403,964,900	22.06	520,605,000	28.42	27,496,600	1.50
1976	1,945,234,800	934,109,100	48.02	425,157,400	21.86	559,241,200	28.75	26,727,100	1.37
1977	†2,056,257,600	986,083,100	†47.95	446,549,500	†21.72	†595,042,900	†28.94	28,582,100	†1.39
1978	†2,123,645,600	981,910,600	†46.24	502,796,600	†23.68	†613,328,400	†28.88	25,610,000	†1.20
1979	2,096,063,300	978,081,200	46.66	491,515,200	23.45	601,079,500	28.68	25,387,400	1.21
1980	1,991,820,900	920,510,700	46.21	509,304,100	25.57	†538,584,000	27.04	23,422,100	1.18

*The amounts carried by motor carriers are estimates. †Revised. ‡Preliminary.

Source: Association of Oil Pipelines, *Oil & Gas Journal*, 13 September 1982, p. 27

oil produced within the United States is consumed within the country, the amount transported by tanker is much less than in many other countries.

At each end of a tanker voyage, ports and terminals serve as the connection between the tanker and land transportation systems. In producing fields, pipelines move crude to the port where it is stored until a tanker is available. When the tanker reaches its destination port, the crude is again transferred to a land transportation system, usually a pipeline. Port and terminal facilities include pumps, storage, offshore loading facilities, control systems, and related facilities.

Not all tankers are loaded at shore-based facilities. Some offshore producing fields include tanker loading capability, and produced crude is piped directly onto a tanker rather than being pipelined to a shore-based terminal for loading. These offshore terminals include an offshore structure extending above the water's surface that transfers crude from storage into the tanker. They are remote from other offshore structures and provide tanker mooring and transfer piping. The mooring system is designed to allow the tanker to move under the forces of waves and wind while being loaded. Such terminals (single point mooring terminals) are also used at tanker destinations where tanker size prohibits movement into a shallow water port.

Several designs of these facilities have been developed for use in offshore producing fields. Many are capable of withstanding the forces of severe ocean environments.

Crude and products tankers are classified as general purpose/product carriers ranging in size up to 25,000 dwt; supertankers and large tankers range from 25,000–150,000 dwt; very large crude carriers (VLCC) range from 150,000–300,000 dwt; and ultra large crude carriers (ULCC) are above 300,000 dwt. Each classification is particularly suited to a specific application. Larger tankers are used for long voyages; smaller vessels are used for shorter hauls and for "lightering" large tankers when the displacement of the larger tanker is too great to enter a port facility. VLCCs, for example, traditionally have moved crude from producing countries in the Middle East and Nigeria to Europe, Japan, and North America.

Some oil tankers are owned by petroleum companies that normally use the vessels to transport the company's own cargoes; some are owned by independent shipping companies; and some are owned by governments. Tankers are registered in maritime countries such as France, Japan, and the United States, which use their own ships for their own use; in countries such as the United Kingdom, Norway, and Greece, which use their ships mainly to carry cargoes among nations other than their own; and in flags-of-convenience countries such as Liberia, Panama, Cyprus, Singapore, and Bermuda. Flags-of-convenience countries do not need most of the capacity of ships registered in the country for their own exports and imports, but the vessels registered there are a source of national income. A flag-of-convenience vessel can be controlled and manned by non-nationals, and there is little or no tax on income from the ship.

Flags of convenience are used extensively because they make transportation arrangements flexible, especially in times of war, boycotts, and other crises. Vessels registered under flags of convenience are often able to compete better in the market because they can offer a lower cost. The more competitive cost results from builders' ability to search for the lowest-cost construction yards and lower manning costs, and from the fact that the vessels are not subject to cargo preference laws.

Oil companies own a large number of the tankers used in crude oil transportation. When needed, additional capacity is supplied by tankers on a long-term charter basis in which a tanker is leased for a specified period. The tanker owner normally furnishes crews, supplies, and insurance; the charterer pays fuel costs, port charges, and miscellaneous costs. Transportation above that supplied by oil company-owned vessels and vessels under long-term charter is supplied by tankers operating on a spot, or single voyage, basis. Spot market transactions account for about 10% of crude tanker movements.[7]

The cost of ocean tanker crude transportation varies. The spot market is particularly dependent on the demand for tankers and on transportation costs. The cost of operating a tanker owned by the oil company includes the operating cost of the vessel, the cost of capital to acquire the vessel, and the cost of operating-related facilities.

The current basis for comparing tanker costs is the Worldscale, developed to replace the International Tanker Nominal Freight Scale issued in London and the American Tanker Rate Schedule issued in New York. The Worldscale Association (New York) now computes base rates for the western area of the world; the Worldscale Association (London) does the same for the rest of the world. Worldscale rate calculations are complex and involve several factors: a standard vessel specification, a fixed-hire element, bunker prices and port costs, brokerage, laytime allowance, port time, and canal transit time. A brief explanation of how Worldscale rates are used for decision making is included in Nersesian's *Ships and Shipping*.

Changes in the tanker fleet, in addition to the entry of producing countries with their own transportation companies, are expected. One study included the following projections for the future tanker fleet:[9]

1. Vessels will still be large, but not as large as some of those existing today. Weights may be restricted to 200,000–250,000 dwt with limited draft to allow the tanker to make use of the Suez Canal.
2. Vessels will be more versatile, perhaps with more, but smaller, tanks.
3. Tankers will be more fuel efficient and slower. Speeds of 12–13 knots rather than 15–16 knots will be common. This will require advanced

power and propulsion systems, possibly even including coal firing as an alternative.

LNG tankers.   Beginning in the 1960s, ocean transport of natural gas in liquid form expanded. Large natural gas reserves in producing countries such as Algeria and Iran far exceeded the market that could be easily reached by pipeline. In countries that produce large amounts of oil, large volumes of "associated" gas are also separated from the oil. Until recent years, much of this associated gas was flared (burned in the producing field) because markets were not available.

Increases in the price of natural gas and gas liquids, coupled with efforts to conserve the gas that was previously flared, gave impetus to a growing international trade in liquefied natural gas (LNG). When natural gas is liquefied by cooling, it is reduced in volume by about 600 to 1, making it economical to transport by special ocean carriers.

The LNG tanker is only one part of the movement of LNG from producing field to market. A liquefaction plant is required near the port where the LNG is loaded aboard the tanker. Natural gas is moved from the producing field to the liquefaction plant by pipeline in the gaseous state. After liquefaction, LNG is loaded onto the tanker through short loading lines connecting the liquefaction plant with offshore loading facilities.

At the tanker's destination in the consuming country, another plant is required to regasify the LNG for distribution to users as natural gas. Storage at both origin and destination facilities is also required.

Investment in LNG carriers, or tankers, can represent as much as one-half the total investment in an LNG project consisting of carriers, a liquefaction plant and related facilities in the exporting country, and a gasification plant in the importing country. One reason for this enormous cost is that special materials and special welding methods must be used in LNG carrier construction. To liquefy natural gas, it must be cooled to about $-260°F$. Conventional steels become brittle at this low temperature. Insulation is also critical in the design and construction of LNG tankers; cargo temperature must be maintained at $-260°F$ to minimize losses from vaporization.

The amount of vaporization (boiloff) during an LNG voyage—with a well-designed vessel and insulation and good operating practice—is typically below 0.25% of the cargo per day.

At the end of the voyage, some LNG is kept on board to vaporize during the return trip to maintain the storage tanks at a low temperature. If the tanks warm up during the trip, loading at the export terminal must be delayed to prevent damage caused by "cold-shocking" the tanks. Also, the fewer temperature cycles to which the tanks are subjected, the longer their life.[10]

Several LNG carrier designs have been developed. The type of cargo tanks and how they are attached to the vessel are key differences among the designs.

The future of LNG tanker transportation is difficult to assess due to political influences and ever-changing economics. But large amounts of natural gas in countries where it cannot be fully utilized—coupled with efforts around the world to conserve energy that has been previously wasted (flared)—should bring growth in international LNG trade. As the price of natural gas continues to rise, the economics of LNG transport will improve. Currently, some key LNG tanker routes and their associated sources and markets include natural gas from Algeria to France, Great Britain, the United States, and Spain, and shipments from Alaska, Brunei, Indonesia, and Abu Dhabi to Japan.

LPG. Another category of ocean petroleum vessel is the liquefied petroleum gas (LPG) carrier. These carriers transport propane and butane in liquid form for use in refining and as a petrochemical feedstock and for direct use by industry. LPG is produced in refineries and is also recovered from natural gas in gas-processing and liquefaction plants.

LPG carriers are not as costly as the more specialized LNG vessels; minimum temperature required is in the $-50°F$. range and more conventional materials and construction techniques can be used.

Large volumes of liquefied petroleum gases are expected to be available in Middle East countries where liquid-hydrocarbon-rich associated gas is produced with oil. Ambitious programs in many producing countries to capture and utilize this gas rather than flare it will increase the volume of gas liquids available.

REFERENCES

1. John N. Hooker, "Oil Pipeline Energy Efficiency Studied for U.S.," *Oil & Gas Journal,* (15 February 1982), p. 114.
2. John P. O'Donnell, "Petroleum 2000: Pipelines Continue to Play Major Role," *Oil & Gas Journal,* (August 1977), p. 279.
3. Earl Seaton, "U.S. Pipelines Keep Energy Moving," *Oil & Gas Journal,* (22 November 1982), p. 73.
4. Earl Seaton and Bob Tippee, "Non-Communist Pipeline Plans Total 49,334 Miles in '83, Beyond," *Oil & Gas Journal,* (24 January 1983), p. 23.
5. Joseph M. Morgan, "OGJ-Morgan Oil Pipeline Cost Index: Recession Trims Third-Quarter Building Costs," *Oil & Gas Journal,* (9 May 1983), p. 119.
6. Interstate Natural Gas Pipeline Companies—1980, United States Department of Energy.
7. Alex Marks, *Elements of Oil Tanker Transportation,* Tulsa: PennWell Publishing Co., 1982.
8. "Tankers Cut Pipelines' U.S. Transportation Share," *Oil & Gas Journal,* (13 September 1982), p. 27.
9. *International Petroleum Encyclopedia,* Tulsa: PennWell Publishing Co., 1981.
10. Roy L. Nersesian, *Ships and Shipping: A Comprehensive Guide,* Tulsa: PennWell Publishing Co., 1981.

2

TYPES OF PIPELINES

MOST oil and gas pipelines fall into one of three groups: gathering, trunk/
transmission, or distribution. Other pipelines are needed in producing
fields to inject gas, water, or other fluids into the formation to improve oil and
gas recovery and to dispose of salt water often produced with oil.

Small-diameter pipelines within an oil or gas field, called flow lines, are
usually owned by the producer. They connect individual oil or gas wells to
central treating, storage, or processing facilities within the field (Fig. 2–1).
Another gathering system made up of larger-diameter lines, normally owned by
a pipeline company rather than the oil or gas producer, connects these field
facilities to the large-diameter, long-distance trunk or transmission line. In some
cases, individual wells are tied directly to the pipeline company's gathering
system.

Crude trunk lines move oil from producing areas to refineries for processing.
Gas transmission lines carry natural gas from producing areas to city utility
companies and other customers. Through distribution networks of small
pipelines and metering facilities, utilities distribute natural gas to commercial,
residential, and industrial users.

Oil pipelines

Flow lines, the first link in the transportation chain from producing well to
consumer, are used to move produced oil from individual wells to a central point
in the field for treating and storage. Flow lines are generally small-diameter
pipelines operating at relatively low pressure. Typical flow-line diameters in the

Fig. 2–1. Flow lines bring individual well streams to gas-oil separation facilities.
Source: *Oil & Gas Journal,* 25 July 1977, p. 153.

United States are 2 in., 3 in., and 4 in. The size required varies according to the capacity of the well being served, the length of the line, and the pressure available at the producing well to force the oil through the line. Flow lines typically operate at pressures below 100 psi. In many fields around the world, high-capacity wells require larger-diameter pipelines.

Individual oil flow lines are relatively short, typically ranging from less than a mile to a few miles. However, an oil field containing many wells, each of which is connected to central facilities by a flow line, can contain several hundred miles of pipeline in a relatively small geographical area.

The throughput of oil flow lines ranges from a few bbl/day upward, depending on the capacity of the well being served. Many wells produce several hundred bbl/day, for instance, and some wells may produce as much as several thousand bbl/day.

Offshore, relatively few flow lines are installed. For economic and operating reasons, most offshore development wells are directionally drilled from central platforms, permitting the wellheads to all be placed in a small area on the platform. Individual wells therefore do not normally need to be connected by undersea pipelines to processing equipment. However, a few hundred offshore wells have been drilled remote from central abovewater platforms and are connected by flow lines laid on the ocean floor. Also, in some cases a platform containing a number of wells will be connected by an undersea line to a separate platform where the oil is processed and/or stored.

Because oil flow lines are short, the energy (pressure) required to move the oil through the pipeline to central facilities within the field is relatively low. There are two types of oil wells: those that flow unaided because of the natural energy of the reservoir and those that must be pumped. The pressure that forces oil in a flowing well to flow to the surface is usually sufficient to move the oil on to the central field facility. In wells in which a bottom hole pump must be used to lift the oil to the surface, the pump's energy also moves the fluid through the flow line. Additional pumps at points along the flow line are not normally needed.

The destination of most oil flow lines is a tank battery. One or more tank batteries may be installed in a single field, each serving a number of individual wells. A typical tank battery contains a separator to separate oil, gas, and water; a fired heater to break water-oil emulsions to promote complete removal of water from the oil; and tanks for storing the oil until it can be shipped from the lease by truck or pipeline. Metering equipment is also included to measure the volume of oil leaving the lease. An additional separator, separate meters, and other equipment may also be installed for periodic testing of individual wells.

The oil in each flow line coming to the tank battery from an individual well is measured before being mixed with the flow from other wells for treating and separation. This information is important for evaluating the performance of the well and the reservoir.

Other equipment may be required at these field facilities under special conditions. Desalting facilities are needed if the produced crude contains large amounts of salt, and heated storage may be required if the oil is too viscous at low temperatures to be pumped from lease storage.

Flow lines are normally made of steel, though various types of plastic pipe have been used in a limited number of applications. Sections, or joints, of steel pipe for flow lines can be connected by welding or by the use of threaded couplings. Other specialty joints and joining methods aimed at reducing construction time and cost have also been developed for both steel and other flow-line materials.

Pipe used for oil flow lines is relatively lightweight because operating pressures are low. Wall thickness for a 3-in. diameter flow line, for example, might typically be 0.216 in., corresponding to a weight of 7.58 lb/ft. Heavier pipe in the 3-in. size is available in wall thickness to 0.437 in. and in weights up to 14.3 lb/ft. Pipeline pipe is usually referred to by its *nominal* size, 3-in. in this case. The outside diameter of nominal 3-in. diameter pipe is actually 3.500 in.

More complete details on pipe weights, grades, and sizes are given in Chapter 3.

Some flow lines are coated internally to protect against corrosion. Whether or not internally coated pipe is used depends on the corrosion potential of the oil, the expected producing life of the well being served, and other factors. Where flow lines are buried, they are usually also coated externally to minimize corrosion.

When water and gas have been removed from the oil, it is stored in lease tanks for shipment. Oil may be trucked from the lease if a pipeline is not available, but this method is used primarily when small volumes of oil are produced on the lease and a pipeline is not justified, or when a new well is completed and the pipeline has not yet been laid to the lease.

Oil leaving the lease must be measured, either manually or automatically. Manual measurement involves *gauging* the lease tanks before and after oil is removed. The volume shipped is then calculated. Oil can be shipped from the lease by manually operating a valve in the storage tank that lets oil flow into a truck or into the pipeline company's gathering line.

Today, lease automatic custody transfer (LACT) units are used where significant oil volumes are involved. In this method, a pump is automatically started when the level in the storage tank reaches a prescribed height, and oil is pumped into the gathering line. The pump remains on until the level in the tank is lowered to a designated point; then the pump is automatically shut off. The volume of oil flowing through the LACT system is automatically measured. A sampler also measures the water and sediment in the stream so a correction can be made to the volume measurement when calculating the payment to the lease owner. In fields producing large volumes of oil, shipment may be virtually continuous from the lease storage tanks.

The next link in the oil pipeline chain is gathering lines that transport oil from field-processing and storage facilities to a large storage tank or tank farm where it is accumulated for pumping into the long-distance crude trunk line. These gathering systems are normally owned by the pipeline company that operates the main trunk line. In the United States, these systems typically consist of lines ranging from 4 in. to 8 in. in diameter. Size, of course, depends on the volume of crude to be moved, pipeline length, and other factors. Operating pressure is higher than that of field flow lines.

Gathering system throughput obviously varies widely, depending on the number of field storage tanks served and the producing capacity of the wells in each field. These gathering systems are quite flexible; their capacity can be increased through various methods to accommodate new producing fields in an area or other volume changes.

The mileage contained in both crude and products-gathering systems in the United States is reported by the FERC to be about 38,500 miles.[1] Though an accurate count of gathering-system mileage outside the United States is not available, the same concept is used in gathering oil production.

Crude trunk lines. From large central storage facilities, oil is moved through large-diameter, long-distance trunk lines to refineries or to other storage terminals (Fig. 2–2). In the United States, much of this traffic is from the oil-producing areas of the West, Southwest, and Gulf Coast to refining centers in the central and upper Midwest and the U.S. Gulf Coast.

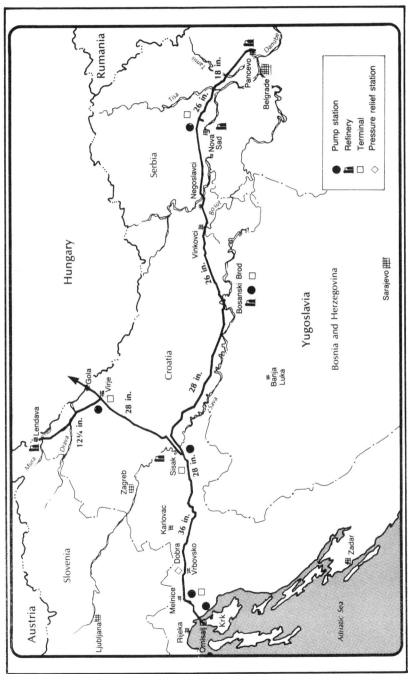

Fig. 2-2. Example crude trunk line system. Source: *Oil & Gas Journal*, 8 February 1982, p. 112.

This network of crude trunk lines comprises a wide variety of pipe sizes and capacities. Pumps are required at the beginning of the trunk line, and pumping stations must also be spaced along the pipeline to maintain pipeline pressure at the level required to overcome friction, changes in elevation, and other losses. The different sections of the system are sized to handle expected volumes; if new fields must be tied in by a new branch line, the capacity can often be increased by installing additional pump stations.

Crude trunk lines operate at higher pressures than field-gathering systems and are also made of steel. Individual sections are joined by welding. These lines are, in the United States at least, almost always buried below ground surface and are coated on the exterior to protect the steel pipe from corrosion.

Crude oil trunk lines serving the United States can be several hundred miles long. Control of such a system is a complex operation. Sophisticated monitoring and control systems have been developed to permit the pipeline operator to fulfill delivery commitments and avoid a malfunction of the system.

The complexity of these systems varies so widely that it is difficult to select a typical system. The fact that they traverse long distances complicates their construction and operation. Flow lines are usually confined to a single field, and the parties involved in the decision making and permitting are few. But when a line must cross land owned by many different owners, most of whom receive no benefit from the pipeline, just the job of obtaining right of way, for example, becomes significant.

Environmental laws also require that many permits be obtained to cross roads and streams, pass through wildlife areas, and for other purposes. For example, about 1,400 permits had to be obtained from various state and federal agencies to begin construction of a crude pipeline from the West Coast to Minnesota.[2]

The trans-Alaska crude pipeline, completed in 1977, is one of the most widely publicized examples of special requirements involved in building a long-distance pipeline. Not only was the permitting process a complex and lengthy one—an act of Congress was required to speed environmental review in the courts—but new techniques and equipment had to be specially developed to ensure the line did not damage the sensitive Alaskan environment. The possibility of a significant oil spill also had to be minimized.

It is difficult to pinpoint a typical throughput for an individual crude trunkline. The trans-Alaska pipeline, a 48-in. diameter line, is designed to carry up to 2 million b/d, but that volume is by no means typical of all systems. However, data are available on an individual company basis for common carrier oil pipelines. Total deliveries of crude by those pipelines was reported by the FERC as about 6.1 billion bbl in 1981.[3] Those companies operated almost 58,000 miles of crude trunk lines.

Gas pipelines

The purpose of gas-gathering pipelines and gas transmission lines is similar to that of crude-gathering and crude trunk lines, respectively, but operating conditions and equipment are quite different. In general, gas pipelines operate at higher pressures than crude lines; gas is moved through a gas pipeline by compressors rather than by pumps; and the path of natural gas to the user is more direct.

Gas gathering. As in the case of oil wells, gas-well flow lines connect individual gas wells to field gas-treating and processing facilities or to branches of a larger gathering system. Most gas wells flow naturally with sufficient pressure to supply the energy needed to force the gas through the gathering line to the processing plant. Downhole pumps are not used in gas wells; but in some very low pressure gas wells, small compressors may be located near the well to boost the pressure in the flow line to a level sufficient to move the gas to the process plant.

Flowing gas well pressures vary over a wide range. At the low end of the range are those for which a compressor must be installed near the well. However, many gas wells produce at such high pressures that pressure must be reduced at the wellhead before the gas enters the flow line. This permits use of lighter-weight, less-expensive steel pipe. Pressure is reduced at the well by a choke or pressure-reducing valve. These can be manually operated or of the type that automatically maintains a prescribed pressure downstream of the valve.

Flow lines from individual wells carry gas to the field processing plant where the gas is treated to make it suitable for sale. Liquid hydrocarbons are also separated from the wellstream for sale. In some cases, several individual well flow lines feed into a larger line, which then carries the combined flow to the plant.

Contracts for the purchase of natural gas from a processing plant by a gas pipeline operator limit the amount of water that can be contained in the gas when it enters the gas transmission line. This limit is normally 7 lb/MMcf (million cubic feet). A dehydration process in the plant removes water to an acceptable level. Also specified in gas purchase contracts is the maximum amount of sulfur the sales gas may contain. If the produced gas is *sour*—contains acid gases hydrogen sulfide or carbon dioxide—these components must be removed in the process plant. Most field gas processing plants also remove hydrocarbon liquids from the produced gas stream. The amount of each component removed varies with the capability and design of the plant, general economic conditions, and market conditions for natural gas liquids and natural gas. But a field gas processing plant typically removes varying amounts of ethane, propane, butanes, and heavier hydrocarbon liquids from the gas stream. A mixture of components heavier than butane is often marketed as one product, natural gasoline.

Water and acid gases are removed from the wellhead gas stream because they can cause corrosion and other problems in long-distance pipelines and associated equipment. Hydrocarbon liquids are removed because of their value as individual products for petrochemical feedstock and other uses.

Lengths of individual gas well flow lines vary, but they are normally from less than a mile to a few miles long. The lines are relatively small; diameters typically range from 2 in. through 4 in. Operating pressures also vary over a wide range but in general are higher than the operating pressures of oil-well flow lines. Gas-well flow lines may operate at several hundred psi, and in some cases up to 2,000 psi or more where plant and field operating conditions make bringing the gas to the plant at a high pressure desirable. Where wells produce at high pressure, this pressure can sometimes be used to provide energy in the gas processing plant. If the pressure were reduced at the wellhead by a choke or pressure-reducing valve, that energy would be dissipated. If, however, the wells flow to the plant at high pressure, that energy can be used within the plant to drive equipment or provide refrigeration for the process.

Length, operating pressure, size, and throughput of gas well flow lines depend on the capacity of the producing well, the type of gas produced, process plant operating conditions, plant location, and other factors.

An accurate measure of the total mileage of gas well flow lines is not available. There are, however, more than 850 natural gas processing plants in the United States alone, and more than 420 such plants exist elsewhere in the world.[4]

Gas transmission. From field-processing facilities, dry, clean natural gas enters the gas transmission-pipeline system for movement to cities where it is distributed to individual businesses, factories, and residences. Distribution to the final users is handled by utilities that take custody of the gas from the gas transmission pipeline and distribute it through small, metered pipelines to individual customers.

Like crude trunk lines, gas transmission systems can cover large geographical areas and can be several hundred miles long or more. Much gas is moved, for example, from Texas and Louisiana to the populated areas of the northeastern United States. A plan to bring natural gas from Alaska's North Slope would involve a gas pipeline system about 4,800 miles long. Work was also underway in 1982 on a large gas line about 2,800 miles long to bring Russian gas to western Europe. Other long-distance gas transmission lines include a system that transports gas from Iran to Russia. In the United States alone, where natural gas transmission pipelines are regulated, the FERC reported that natural gas pipeline companies in 1981 operated about 73,000 miles of field-gathering lines and 197,000 miles of transmission lines.

Gas transmission lines operate at relatively high pressures. Compressors at the beginning of the line provide the energy to move the gas through the pipeline. Then compressor stations are required at a number of points along the line to maintain the required pressure (Fig. 2–3). The distance between compressors varies, depending on the volume of gas, the line size, and other factors. Capacity of the system can often be increased by adding compressors at one or more of these compressor stations or by building an additional compressor station. The size of compressors within the station varies over a wide range, but many stations include several thousand horsepower in one station.

Gas transmission pipelines are made of steel pipe and are buried below ground surface. The individual sections of pipe are joined by welding, and the pipe is externally coated to protect against corrosion. Pipe size ranges up to 42

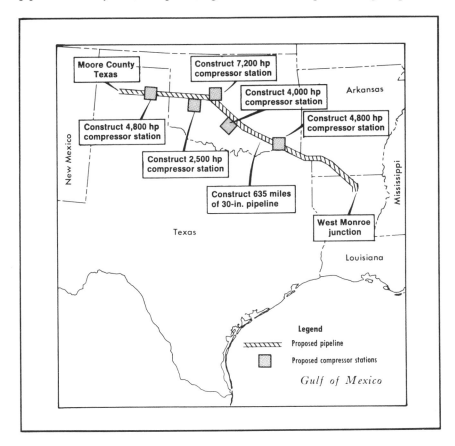

Fig. 2–3.   Portion of gas pipeline system. Source: *Oil & Gas Journal,* 14 July 1980, p. 155.

in. in the United States, and pipe as large as 60 in. in diameter has been installed in the USSR.

Volumes handled by individual systems, as in the case with crude trunk lines, vary widely. But in 1981, the FERC reported that United States natural gas pipeline sales amounted to 18.5 trillion cu ft.

The operation of a gas transmission system that moves gas over a large geographic area and contains several compressor stations and other facilities is a complex control challenge. Computers and sophisticated communications systems have been joined to allow pipeline operators to deliver the volumes required and to minimize malfunctions of the system. Because the needs of customers change more frequently and more rapidly, control of natural gas pipeline deliveries can be even more complex than the operation of a crude trunk line. The effects of a lack of natural gas, because gas provides home heating and fuel for business and industry, can be felt more immediately in some cases than a disruption in the delivery of crude to a refinery.

Alberta Gas Trunk Line (now NOVA, An Alberta Corporation) is an example of a large gas-gathering and transmission system. NOVA gathers and transports all of the natural gas in the province of Alberta, Canada, that is to leave the province. In 1979, NOVA operated over 6,500 miles of gas pipeline, ranging in size from 3-in. diameter to 42-in. diameter. The system had over 500 gathering points and 32 compressor stations.[5] Total compressor horsepower was about 500,000 hp, and throughput was about 5 billion cu ft/day.

Products pipelines

The industry's products pipeline system, especially in the United States, is a sophisticated transportation network. Many segments of the system are highly flexible in both capacity and the products that can be transported.

One part of this system moves refined petroleum products from refineries to storage and distribution terminals in consuming areas. Products shipped include the several grades of gasoline, aviation gasoline, diesel, and home heating oils. In the United States, much of this movement is from Gulf Coast refining centers to the East and Southeast. But significant volumes of these products also are shipped from the Gulf Coast to the upper Midwest. In other countries products pipelines may move refined products from coastal refineries or tanker unloading terminals to the interior of the country to supply populated areas.

Another group of products pipelines is used to transport liquefied petroleum gases (LPG) and natural gas liquids (NGL) from processing plants in oil and gas-producing areas to refineries and petrochemical plants (Fig. 2–4). In some cases, a mixed stream of liquid hydrocarbons separated from natural gas at field processing plants is moved to a fractionation plant where the mixed stream is

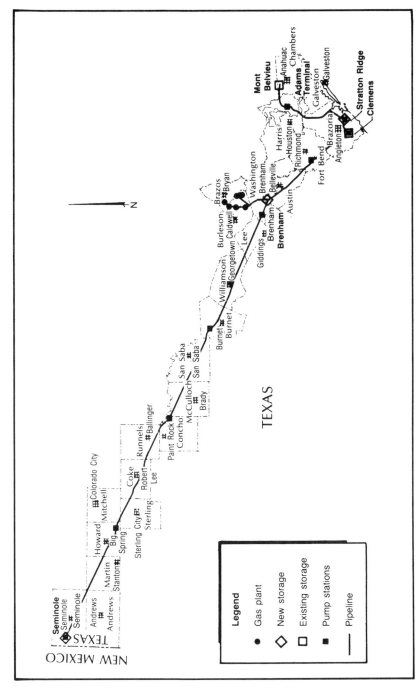

Fig. 2–4.   Example products pipeline system.   Source: *Oil & Gas Journal*, 14 September 1981, p. 102.

separated (*fractionated*) into individual products, including ethane, propane, and butanes.

Products pipelines can often carry several different products in the same pipeline. Though there is a short length of the pipeline in which two such "batched" products may be mixed, operating methods allow the purity of each product to be maintained. Batching is done either with or without a physical barrier separating the two products. Where no physical barrier is used between different products, the difference in density of the two materials maintains the separation (under pressure and in turbulent flow) with only a short length interval in which mixing occurs. The position of each batch and the extent of mixing can be monitored at points along the line by measuring the density of the fluid in the line. Sphere batching is also used. A sphere can be inserted in the pipeline to form a physical barrier between batches of different products to maintain separation.

Movement of more than one product in a single pipeline obviously calls for even more sophisticated monitoring and control than is required for continuous movement of a single product.

Products pipelines often must operate at higher pressures than crude pipelines because the material being transported is lighter than crude. Products being shipped must remain in a liquid phase rather than become a mixture of gas and liquid. If gas is allowed to enter the liquid pumps on the pipeline, pump efficiency is lowered and pump damage may result. In general, lighter (lower-density) materials require higher operating pressures to prevent formation of gas in the pipeline. For instance, one products pipeline that moves ethane from ethane extraction facilities to ethylene manufacturing facilities and underground storage sites has a maximum operating pressure of 1,440 psi. To prevent vaporization, design criteria for this pipeline called for a minimum pump suction pressure of 650 psi. Ethane is the lightest hydrocarbon transported in products pipelines; the pressure at which vaporization occurs decreases as the density of the material being shipped increases.

Like crude and natural gas pipelines, there are no typical products pipelines or products pipeline systems, but 8-in. diameter through 16-in. diameter products pipelines are common. These are certainly not limits on pipeline size; both smaller and larger lines are in service. In Saudi Arabia, for instance, a 730-mile-long pipeline to move natural gas liquids from field processing plants in the eastern province to the west coast consists of 26-in. and 28-in. diameter pipe.[6] Throughputs and capacities of products pipelines not only differ among systems, but the capacity of a single line varies with the material being shipped. And the capacity of a given line for a given product can often be increased by installing additional pumps at pump stations. As an example—it is by no means average—one products pipeline system that moves LPG, diesel, gasoline, and heating oil from a refinery in West Texas to a distribution center in eastern Kansas consists of 16-in., 12-in., and 8-in. lines and has a capacity of 140,000 b/d.[7]

Another system shows the complexity and flexibility of a modern products-pipeline system. It connects gas processing plants in New Mexico, Colorado, Wyoming, and Utah to a pipeline and distribution system in the upper Midwest. The 1,170 miles of main line and branch gathering lines carries a mixed stream of ethane, butane, propane, and natural gasoline. Operating pressure of the line is 1,600 psi. Initial capacity was to be 35,000 b/d, but eventually the system is expected to handle 65,000 b/d.[8] The main line of the system is 12-in. and 10-in. diameter pipe, and laterals are 8-in., 6-in., and 4-in. diameter.

Initially, the system would have five pumping stations with a total of 8,000 operating pump horsepower. When capacity is boosted to 65,000 b/d, five additional stations would be installed and total pump horsepower would be 20,000 hp. This project also provides an example of what is involved in crossing wildlife areas, just one of the many types of areas for which permits must be obtained and in which activity is restricted.

In the United States, pipeline firms reporting to the FERC operated over 76,000 miles of products pipelines in 1981. Product deliveries by these companies through this network amounted to more than 4.15 billion bbl during that year.[9]

Other pipelines

In addition to oil, gas, and products pipelines, there are other types of energy-related pipelines. Many are operated by oil and gas producers or groups of producers to solve special problems, to increase the production of oil and gas, or to market traditional products in a more efficient way. Some of these pipelines carry oil and gas in somewhat different forms or combinations. Multiphase pipelines, for example, carry oil and gas together in the same line at the same time; LNG pipelines carry natural gas, but in a liquid form rather than as a gas in the case of gas transmission lines.

Pipelines also carry nonpetroleum products. Coal slurry lines, for example, may become a much more significant factor in energy transportation in the years to come, especially in the United States where large coal reserves exist. And the industry's emphasis on enhanced recovery of oil in the United States in recent years has resulted in the design and construction of pipelines to carry carbon dioxide to oil-producing areas for injection into the oil reservoir.

Two-phase pipelines. In most cases, it is desirable to transport petroleum as either a gas or a liquid in a pipeline. In a line designed to carry a liquid, the presence of gas can reduce flow and pumping efficiency; in a gas pipeline, the presence of liquids can reduce flow efficiency and damage gas compressors and other equipment. Different materials can be moved in a crude or products pipelines as discussed earlier, but both are normally transported as liquids.

There are, however, cases where it is more economical or more practical to transport both liquid and gas in the same pipeline at the same time. In general, these situations occur when the volumes to be moved are relatively small and flow efficiency is not a critical factor, or where the construction of two separate lines—one for liquid and one for gas—will be very expensive.

An example of the first situation is a flow line that brings production from an individual well to a field processing facility. Most oil wells produce some natural gas along with the crude. The amount of gas produced with the oil—the gas-oil ratio—varies from field to field and often changes with time in a single well.

Many gas wells also produce large amounts of liquids, or condensate. In these cases both oil and gas produced from a single well can usually be accommodated in one flow line to the field's crude separation or gas-processing facility. The flow line usually has more than adequate capacity, and flow efficiency is not critical; the distance the fluids must travel is also relatively short.

Two-phase pipelines have also been used offshore where pipeline construction costs are high. In this case flow efficiency in the line may be very important, but careful design of the system can make a two-phase line an economic alternative to the construction of two separate pipelines.

The design of pipelines for two-phase flow is a complex problem. Continued research over decades has failed to simplify the problem to any great extent. In any pipeline that must handle a specified throughput, an estimate of the pressure drop in the line is a key to proper design of line size and pump capacity. This estimate is difficult in the case of two-phase pipelines because there are several flow regimes—laminar, plug, slug, turbulent—that can be present in the line. The pressure drop that must be overcome by pumps is different for each of these regimes, and it is difficult to tell which of the regimes will be present for a given set of operating conditions. A slight change in flow or pressure can often move the flowing conditions into another regime, drastically affecting the efficiency of the pipeline. Additional equipment is also often needed in a two-phase pipeline system to handle liquid "slugs."

These uncertainties have limited the use of two-phase pipelines to those applications where there is no other reasonable alternative. But such a pipeline system can be designed and operated satisfactorily. For example, two 32-in. diameter, 360-km-long two-phase pipelines are being successfully operated in the North Sea, and more such facilities are planned. Computer modelling of two-phase flow conditions is being developed as a way to arrive at a more accurate design, and experimental work is being done to confirm theoretical data.[10] The design of a two-phase pipeline requires more accurate data than the design of a single-phase pipeline; even the route can have a significant effect on the satisfactory operation of a two-phase pipeline.

LNG pipelines. Liquefied natural gas (LNG) is natural gas cooled and compressed to a temperature and pressure at which it exists as a liquid. Significant volumes of natural gas are transported in the liquid phase as LNG, but these shipments are made by special ocean tanker rather than by long-distance pipeline.

Short pipelines, however, are in operation in association with gas liquefaction or vaporization plants and terminals for loading and unloading of LNG tankers. Natural gas is moved from the producing fields in the gaseous form to a liquefaction plant near a shipping port where the gas is liquefied for loading aboard an LNG tanker. At the tanker's destination in the consuming area, LNG is unloaded from the tanker to storage, and the LNG is again vaporized for distribution in the gaseous state to consumers. Liquefaction of natural gas provides a way to transport it from producing countries to consuming countries when natural gas pipelines are not possible. For land transportation, pipelining natural gas in the vapor phase is still the preferred method.

But the feasibility of long-distance LNG pipelines has been studied.[11] The key advantage of moving natural gas as a liquid is that as a liquid it has a much higher density. As a result, a smaller-diameter pipeline can be used to transport an equal amount of gas, and less pumping horsepower is required. This ability to carry more gas in less space is one reason for the development of LNG tankers.

The disadvantages of long-distance LNG pipelines stem from the fact that the gas must be kept at a low temperature to maintain it in a liquid phase. This requires insulation of the pipeline and cooling stations to remove the heat that is added by pumping. Special steels will also be required because of the low operating temperatures of an LNG pipeline. An LNG pipeline will also be harder to start up after a shutdown than a vapor-phase pipeline or a crude oil pipeline; it would also be less suitable for operation at partial loading.

There may, however, be some applications in which this approach to moving natural gas long distances has merit. A study by Stuchly and Walker assessed the technology of LNG pipelines and considered a hypothetical pipeline in Canada to carry a design volume of 2,500 MMcfd. That study was based on a 1,430-mile pipeline. Economic and engineering calculations resulted in the selection of a 36-in. pipeline insulated with 6 in. of urethane foam insulation that would operate at a temperature of -140 °F. Minimum operating pressure was set at 540 psia, and maximum operating pressure at 790 psia.[11]

Using the allowable pressure drop and temperatures required to keep the LNG in a liquid phase—and other parameters—the spacing of pumping stations and cooling stations was selected. It was determined that locating the cooling stations at pump station sites would be the most practical and economical approach. The result was a design calling for 20 pumping/cooling stations ranging from 32 to 107 miles apart.

CO_2 *pipelines.* Interest in increasing oil recovery from United States reservoirs has centered on several enhanced oil recovery (EOR) techniques, including the injection of carbon dioxide (CO_2) into the reservoir. Sources of CO_2 include flue gas and natural reserves. In both cases, pipelines are needed to move the CO_2 to the oil-producing field for injection.

A CO_2 pipeline system for enhanced oil recovery will typically include gathering lines to collect CO_2 from the source (wells or other sources) and move it to processing facilities needed before the CO_2 is shipped in a long-distance pipeline; a trunk line to transport the CO_2 from processing facilities near the source to the oil-producing field in which it will be injected; and distribution lines within the oil field through which the CO_2 is injected into individual injection wells.

One large project involves a 495-mile carbon dioxide pipeline to transport naturally occurring CO_2 from Colorado to Texas. The project would include a 30-in. diameter pipeline and was estimated in 1981 to cost about $1.6 billion. Of this, the pipeline and associated compressor stations would cost about $250 million. The remainder would be required for developing the CO_2 supply area, including processing facilities, completing wells, field facilities, and a gas treating plant at the pipeline's oil field destination.[12] The owners estimate that CO_2 injection could recover an additional 280 million bbl of oil from the field, extending its productive life by 20–25 years.

Design considerations for a CO_2 pipeline are unique, but the successful operation of a CO_2 pipeline in West Texas for 10 years has demonstrated it is possible to build and operate such pipelines safely.[13] That pipeline, 180 miles long and 16 in. in diameter, was designed for a maximum operating pressure of 2,035 psi. Adequate operating pressures are necessary to maintain the CO_2 as a liquid. An important design consideration is to provide adequate pressure at pump suction to prevent pump *cavitation.* Cavitation occurs when some of the liquid in the suction line or pump vaporizes; then the vapor bubbles collapse in a higher-pressure region of the pump. It is also necessary to determine accurately any impurities in the carbon dioxide since they have a dramatic effect on the vapor pressure of the CO_2—the pressure below which it vaporizes.

Selection of pipe for CO_2 pipelines is a critical design step because carbon dioxide's unique properties can cause fracture effects that differ from those found in natural gas pipelines.

Operation of a CO_2 pipeline is not hazardous. Carbon dioxide is not toxic even in relatively large concentrations, though its high density (heavier than air) causes it to stay near the ground, increasing the danger of asphyxiation. The danger of a fire from a line break is almost nonexistent.

The total mileage of carbon dioxide pipelines in service is a small share of petroleum industry pipelines. But considerable growth is expected as the use of CO_2 in enhanced oil recovery becomes more widespread.

Coal slurry pipelines. Slurry pipeline systems have been operating for years, carrying finely ground solids in water.

Long-distance coal slurry pipelines are unique in their design and operation. But one large system is moving ahead, and several others are in various stages of planning. Two key obstacles have slowed the development of a coal slurry pipeline network in the United States: a satisfactory solution to problems involved in obtaining large amounts of water; and difficulty in obtaining permission to cross railroad rights of way. Eminent domain legislation required to solve the right-of-way problem was being processed in early 1983, and industry sources gave the bill a good chance of passage.

The water supply situation is aggravated by the fact that the nation's western coal is located in an area where water is especially scarce. But the builder of one major coal slurry pipeline, the Energy Transportation Systems Inc. (ETSI) line, has shown that the water problem is not insurmountable. It has worked out a water transfer agreement that could be applied to other systems. Though the water issue has been an emotional one, water required for coal slurry-pipeline operations is relatively small, compared with many energy processes.[14]

The Black Mesa coal slurry pipeline began commercial operation in 1970 and by late 1980 had shipped about 32 million tons of coal from a mine near Kayenta, Arizona, to a power-generating plant in southern Nevada.[15] Most of the 273 miles of pipeline is 18-in. diameter. The slurry preparation plant includes cage mills for dry crushing, rod mills for wet grinding, and safety screens that limit the particle size that enters the pipeline. After screening, the slurry is diluted to the concentration required for pipelining; the slurry varies from 46–48% solids.

Pump stations have piston pumps. Three stations operate in the 500–800-psi range, while one station operates between 1,250 and 1,650 psi because a high elevation lift is required. At the power generation plant, the slurry is dewatered and the coal stored.

In early 1981, it was estimated that coal slurry pipelines in the United States involving 10,790 miles of pipeline were planned. In addition to ETSI's 1,664-mile, 38-in. diameter line that would carry coal from Wyoming's Powder River basin to Oklahoma, Arkansas, and Louisiana, the following projects were being considered:[16]

1. A 1,500-mile line to move coal from Indiana, Illinois, Kentucky, West Virginia, and Pennsylvania to Georgia and Florida.
2. A line to move Wyoming coal to Minnesota, Wisconsin, Illinois, and Iowa.

3. A 1,100-mile, 20-in. and 24-in. diameter line from Wyoming to Oregon and Washington.
4. A 253-mile, 12-in. and 22-in. diameter line from Utah to Nevada.
5. A proposed 900-mile, 24-in. pipeline to carry coal from Colorado to Houston.
6. A proposed 1,300-mile, 20-in. and 38-in. diameter line from Montana to Houston.
7. About 1,500 miles of 36-in. diameter coal slurry pipeline to move eastern coal to Georgia and Florida.
8. A proposed 650-mile, 26-in. diameter network to carry coal from Utah to southern California.

In an attempt to solve problems involved with using water as a carrier fluid, other carrier fluids have been proposed. These fluids include oil, methanol, and liquid carbon dioxide. Methanol and liquid CO_2 offer the advantage that each can be produced from coal and both are less viscous than water, allowing the movement of larger coal volumes. But methanol is a highly volatile and flammable liquid, and to pump carbon dioxide as a liquid requires higher operating pressures.[17]

The design of a coal slurry system is a unique challenge. In addition to obtaining a source of water, attention must be paid to the size of the coal particles to be transported, the disposition of fines generated as the coal moves through the pipeline, special pumping requirements, and prevention of erosion of the steel pipe.

Capacities of many proposed coal slurry pipelines have not been determined yet, but the ETSI line is planned to carry about 30 million ton/year to power plants along its route. Powder River Pipeline Inc. was also considering a proposal for a system with a capacity of about 36 million ton/year.[18] The 1,923-mile line was estimated to cost $1.26 billion in 1978 dollars.

REFERENCES

1. Earl Seaton, "U.S. Pipelines Keep Energy Moving," *Oil & Gas Journal,* (22 November 1982), p. 73.
2. "Northern Tier Denied State Approval," *Oil & Gas Journal,* (19 April 1982), p. 65.
3. See reference 1 above.
4. Ailleen Cantrell, "Worldwide Gas Processing," *Oil & Gas Journal,* (19 July 1982), p. 103.
5. Ralph C. Hesje, "AGTL Program Boosts Energy Efficiency in Gas Transmission," *Oil & Gas Journal,* (23 July 1979), p. 43.
6. "Work Pressed on Big Lines Across Arabia to Red Sea," *Oil & Gas Journal,* (9 July 1979), p. 116.

7. "Phillips Completes First Segment of Products Line," *Oil & Gas Journal*, (15 September 1980), p. 114.
8. "Mapco's 1,170-Mile Line Will Extend U.S. NGL Arteries," *Oil & Gas Journal*, (16 June 1980), p. 95.
9. Earl Seaton, "U.S. Pipelines Keep Energy Moving," *Oil & Gas Journal*, (22 November 1982), p. 73.
10. J. Corteville, et al., "Two-Phase Flow Key to Offshore Line Design," *Oil & Gas Journal*, (10 August 1981), p. 71.
11. J.M. Stuchly, and G. Walker, "LNG Pipeline Design—1: LNG Long-Distance Pipelines—A Technology Assessment," *Oil & Gas Journal*, (16 April 1979), p. 59; J. M. Stuchly and G. Walker, "LNG Pipeline Design—2: Hydraulics—a Key to Optimizing LNG Pipeline," *Oil & Gas Journal*, (23 April 1979), p. 68; J. M. Stuchly and G. Walker, "LNG Pipeline Design—3: Station Coordination Critical in LNG Pipeline Efficiency," *Oil & Gas Journal*, (30 April 1979), p. 239.
12. "Shell Eyes Construction Start for CO_2 Pipelines," *Oil & Gas Journal*, (25 May 1981), p. 123.
13. Graeme G. King, "CO_2 Pipeline Design—1: Here are Key Design Considerations for CO_2 Pipelines," *Oil & Gas Journal*, (27 September 1982), p. 219.
14. Peter E. Snoek, "Commercial Success of Slurry Pipelines Creates Opportunities for New Applications in the '80s," *Oil & Gas Journal*, (15 March 1982), p. 93.
15. J.G. Montfort, "Operating Experience is Described for Black Mesa Coal-Slurry Pipeline," *Oil & Gas Journal*, (27 July 1981), p. 192.
16. "Proposed U.S. Coal Slurry Line Projects Total 10,790 Miles," *Oil & Gas Journal*, (19 January 1981), p. 29.
17. Snoek, see reference 14 above.
18. "Powder River-Midwest Slurry Line Proposed," *Oil & Gas Journal*, (19 January 1981), p. 41.

3

PIPE MANUFACTURE AND COATING

S TEEL pipe used in pipeline construction is commonly called line pipe to distinguish it from steel casing and tubing, installed below ground in oil and gas wells, and drill pipe, used for oil and gas well drilling. Line pipe differs from other oil country tubular products in that an exterior coating is usually applied to minimize corrosion. The proper design and application of pipe coating is a key to pipeline safety and dependability.

Line pipe comes in a wide range of sizes. It is made from steels with various chemical compositions and different physical properties using several manufacturing processes. The physical and chemical properties of steel used to make line pipe and the manufacturing processes are rigidly controlled to meet the applicable specifications. Specifications also cover pipe dimensions, allowable tolerances, permissible defects, and testing.

Pipe manufacture

Much of the line pipe used today is manufactured according to specifications of the American Petroleum Institute (API). Qualified manufacturers are permitted to use the API monogram on pipe they sell. The marking is a warranty that the manufacturer has obtained a license to use the monogram and that the product that bears the marking conforms to the applicable API specification. The API, however, does not warrant products that bear the monogram.

Though much of the line pipe used in the petroleum industry is manufactured under one of several API specs, the specifications state, "These specifications are not intended to inhibit purchasers and producers from purchasing or

46

producing products made to specifications other than API . . . ." Modifications of certain parts of the specs can also be made when requested by the purchaser.

Key API specs applicable to line pipe include the following:[1]

1. API Spec 5L covers seamless and longitudinally welded steel pipe in Grades A and B.
2. API Spec 5LX applies to high-test line pipe (both seamless and longitudinally welded) in Grades X42 through X70.
3. API Spec 5LU (tentative) covers ultrahigh-test, heat-treated seamless and welded pipe in grades U80 and U100.
4. API Spec 5LS is applicable to spiral-weld line pipe in Grades A and B, and X42 through X70.

API line-pipe grades are designated by their minimum yield strength in pounds per square inch (psi). Yield strength is the tensile stress required to produce a specified total, permanent elongation in a test sample of the steel; the test sample and procedure are detailed in specifications. Grade A line pipe has a minimum yield strength of 30,000 psi; Grade B a minimum yield of 35,000 psi. In the remaining grades, X42 indicates pipe made of steel with 42,000-psi minimum yield strength; X60 pipe has a minimum yield strength of 60,000 psi, etc.

Line pipe manufactured under API specifications is made from open-hearth, electric-furnace, or basic-oxygen steel.

Manufacturing processes. Two general types of line pipe are manufactured: seamless and welded. These designations refer to how each length, or joint, of pipe is manufactured, not how the joints are connected in the field to form a continuous pipeline. Seamless steel pipe is made without a longitudinal weld by hot working lengths of steel to produce pipe of the desired size and properties.

In the welded category, there are several manufacturing processes. They differ both by the number of longitudinal weld seams in the pipe and the type of welding equipment used. Electric-welded pipe is defined by API as "pipe having one longitudinal seam formed by electric flash welding, electric resistance welding, or electric induction welding without the addition of extraneous metal."

Submerged arc welded pipe is "pipe having one longitudinal seam formed by automatic submerged arc welding. At least one pass shall be made on the inside and at least one pass on the outside."

Gas metal arc welded pipe is "pipe having one longitudinal seam formed by continuous gas metal arc welding. At least one pass shall be made on the inside and at least one pass on the outside of the pipe. Gas metal arc welding is an arc-welding process wherein coalescence is produced by heating with an arc between continuous filler metal (consumable) electrode and the work. Shielding is obtained entirely from an externally supplied gas or gas mixture. The shielding gas protects the fluid weld metal from oxidation or contamination by the surrounding atmosphere."

Double-seam welded pipe, specified in API Spec 5L as applicable to Grades A and B in sizes larger than 36 in. outside diameter (OD), has two longitudinal seams formed by the submerged arc welding process or the gas metal arc welding process. The seams are located about 180° apart.

Butt-welded pipe has one longitudinal seam formed by mechanical pressure to make a welded joint; the edges are furnace heated to the welding temperature prior to welding.

API Spec 5LX also describes a process using both gas metal arc and submerged arc welding. Pipe made this way has one longitudinal seam formed by first using gas metal arc followed by submerged arc welding.

Spiral weld pipe has a spiral seam along its length formed by either the electric welding process or the automatic submerged-arc process. At least one welding pass is made on the inside and at least one on the outside of the pipe.

Most of the pipe used for oil and gas pipelines, particularly in the United States, is either seamless or longitudinally welded pipe. But spiral weld pipe has been used increasingly in oil and gas service in many areas of the world. Spiral weld pipe still is used for only a relatively small share of oil and gas pipelines. One reason is that only a small number of mills are capable of producing it to meet specifications. Spiral weld pipe can offer advantages. It can be produced in diameters of more than 64-in. OD, and it is not necessary to use a large number of forming tools for various sizes.[2] The diameter of spiral weld pipe is continuously adjustable so any diameter can be produced from a base material of constant width.

Pipe furnished to API Spec 5L, 5LX, and 5LS may be heat treated using one of several processes: rolled, normalized, normalized and tempered, quenched and tempered, subcritically stress-relieved, or subcritically age-hardened. Heat-treating processes are used to modify the steel's characteristics to give it specific physical properties. Spec 5LU for ultra-high-test line pipe specifies that pipe furnished to that specification shall be heat treated; welded pipe is to be heat treated after welding. Heat treatment must consist of either quenching and tempering, normalizing and tempering, or precipitation hardening.

Chemical properties. The chemical composition of steels is varied to provide specific properties. API specifications give a detailed listing of the amount of each element that can be contained in a given grade of steel used for line pipe.

Carbon is a key component in all steels. The amount of carbon affects the strength, ductility, and other physical properties of steel. Maximum carbon content ranges from 0.21–0.31%, depending on the grade of steel used and the method of pipe manufacture. Maximum and minimum contents are also prescribed in API specifications for various pipe grades for manganese, phosphorous, sulfur, silicon, columbium, vanadium, and titanium. Not all of these materials are present in all grades; some are added to certain line pipe steel grades to provide special properties.

In general, the amount of manganese required in line pipe steel increases as the grade (strength) increases. For instance, the maximum manganese in Grade A pipe is 0.90% and the maximum content in Grade X70 is 1.60% (API specs).

Physical properties. Just as close a watch is kept on physical properties during pipe manufacture as is kept on chemical composition. Tests required on the steel, and on the welded or seamless pipe, include several types of tensile tests, fracture toughness tests, bending tests, ductility tests, and others. The appropriate specifications detail where test specimens will be taken and how the test will be conducted.

In addition, a hydrostatic test must be made on each length of pipe at the mill. Test pressures are outlined in the specification for each grade, weight, and size of pipe. In general, the required hydrostatic test pressure increases with increasing strength (grade) and with increasing wall thickness (weight). For instance, in Spec 5LX, the minimum test pressure for 8⅝-in. OD, Grade X42 pipe with a 0.188-in. wall thickness is 1,370 psi. Pipe of the same size and wall thickness, but in Grade X70, must be tested to 2,290 psi. Pipe of the same X42 grade but with a heavier wall thickness of 0.438 in. must be tested to 3,000 psi. This test pressure of 3,000 psi is the maximum required for any grade or any wall thickness in the 8⅝-in. size under specification 5LX.

Test pressures required for other pipe sizes follow the same trends— increasing with grade and weight with a maximum test pressure that applies to several of the heaviest weights in that size. These hydrostatic pressures are only for testing at the pipe mill. They are not design pressures and do not necessarily have any direct relationship to the working pressure of the pipeline after it is installed. The hydrostatic pressure test performed on the installed pipeline will be determined based on operating conditions and test pressures specified by regulatory agencies and other specifications.

API specifications also prescribe dimensions, weights, and lengths for each size and grade as well as permissible tolerances on these dimensions. Diameter, wall thickness, weight, length, and straightness, for example, are specified, as well as the permissible use of *jointers*. Jointers are two pieces coupled or welded together to make a standard length of pipe. They may be included only up to a specified maximum percent in any order, and the length of each piece must be greater than a specified minimum.

Line pipe manufacturing specifications discuss a number of defects that may cause the pipe to be rejected, including dents, improperly aligned edges at the welded seam, an out-of-line weld bead, improper weld bead height, and excessively hard spots in the steel. Cracks, leaks, laminations, and arc burns are also considered defects. An arc burn is a point of surface melting caused by arcing between the welding electrode and the pipe surface.

Not only are these defects described in detail, but the steps required to repair them are also set forth in the specifications.

Pipe ends.   Pipe is furnished by the manufacturer with either plain ends for welding or threaded ends that will be joined by a threaded coupling. Bell and spigot ends are also furnished in a few of the lighter pipe weights. On pipe furnished with a threaded end, the coupling is screwed onto one end of the pipe and a thread protector is installed on the other end. Threads on the pipe and threads on the coupling must meet specification requirements and tolerances. Unless ordered otherwise, the threaded coupling is screwed on only "hand tight" so that it may be easily removed at the job site for thread inspection and cleaning and the application of thread compound.

Plain end pipe is used for a pipeline in which the individual joints will be welded together. It is furnished with a square-cut end or a bevelled end, depending on pipe size and wall thickness.

If required, pipe ends can also be prepared for special couplings. Some of these couplings require a groove around the circumference of the pipe near each end, for instance.

Marking.   Pipe manufactured under API specifications must be properly marked so pertinent information can be obtained at a glance. Markings include the following:

1. Manufacturer's name or mark
2. The API monogram
3. Size of the pipe in inches
4. Weight of the pipe in pounds per foot
5. The pipe grade (A = Grade A; B = Grade B; X52 = Grade X52; etc.)
6. The process used to manufacture the pipe (S = seamless; E = welded pipe, except butt-welded; F = butt-welded pipe)
7. Type of steel (E = electric furnace steel; R = rephosphorized steel)
8. Heat treatment performed (HN = normalized, or normalized and tempered; HS = subcritical stress relieved; HA = subcritical age-hardened)
9. Test pressure, if higher than the pressure tabulated in the appropriate specification
10. Any other requirements

As an example, the stencil marking for 14-in., 54.57-lb/ft, Grade B, seamless open-hearth, regular weight, plain end pipe is:

AB CO API 14.00 54.57 B S

What's used where. Pipe manufactured by each of the different processes is used in a variety of applications. In general, seamless pipe is manufactured in the smaller sizes because the process is not practical for very large-diameter pipe. Welded pipe is manufacturered in a wide range of sizes. The use of spiral-weld pipe for large-diameter pipelines is growing, in part because of advantages spiral welding offers in the manufacturing process.

The size pipe (diameter) used in any pipeline depends primarily on the volume to be handled. The size selected for a given throughput represents the most economical combination of pumping or compression horsepower and working pressure. For example, a given volume of gas could be transported through a relatively small-diameter pipeline by operating the pipeline at a high pressure and using a large amount of compression horsepower. But this is usually not the most economical design, either from a capital cost or an operating cost standpoint. A larger pipeline could handle the desired volume at lower operating pressure using less compression horsepower. Compressor capital costs and operating costs would be lower, and safety would likely be increased.

In addition, a significantly lower pressure would require a lighter pipe (lesser wall thickness) and would eliminate the need for high-pressure valves and other special equipment. The incremental cost for laying the larger pipeline would be small, relative to the total construction cost. Of course, larger-diameter pipe is more costly, but the added cost could easily be offset by its advantages.

Oil and gas pipeline sizes vary from 2 in. to 60 in. in diameter, depending on the system and required throughput. Typically, flow lines in an oil- or gas-producing field range in size from 2 in. to 6 in. OD; gathering systems consist of pipe ranging from 4 in. to 12 in. in diameter, and long-distance crude trunk lines and natural gas transmission lines can range up to 56 in. in diameter or more. API Spec 5L includes dimensions, weights, and test pressures for plain-end line pipe in sizes up to 64 in. diameter. The ranges given here for different types of pipelines are not limits, but represent typical sizes used in each type of system.

Several weights are available in each line pipe diameter. The weight of the pipe in lb/ft in turn varies as the wall thickness for a given outside diameter. For instance, API Spec 5L, in its table of weights, dimensions, and test pressures for plain-end pipe, lists 22 different weights in the 16-in. diameter size (five weights are special weights), ranging from 31.75 lb/ft to 196.91 lb/ft (Table 3–1). The corresponding wall thickness ranges from 0.188 in. to 1.250 in. As the wall thickness increases for a given outside diameter, the inside diameter of the pipe

TABLE 3–1
Line Pipe Specifications

	1		2		3		4		5		6		7		8	
	Outside Diameter, D		Plain-End Weight, w_{pe}		Wall Thickness, t		Inside Diameter, d		Test Pressure, min.							
									Grade A		Grade A		Grade B		Grade B	
									Std		Alt.		Std		Alt.	
in.	mm	lb/ft	kg/m	in.	mm	in.	mm	psi	100 kPa	psi	100 kPa	psi	100 kPa	psi	100 kPa
12¾	323.9	73.15	109.18	0.562	14.3	11.626	295.3	1590	110	1980	137	1850	128	2310	160
12¾	323.9	80.93	120.76	0.625	15.9	11.500	292.1	1760	122	2210	152	2060	142	2570	177
12¾	323.9	88.63	132.23	0.688	17.5	11.374	288.9	1940	134	2430	168	2270	156	2800	193
12¾	323.9	96.12	143.56	0.750	19.1	11.250	285.7	2120	146	2650	183	2470	171	2800	193
12¾	323.9	103.53	154.08	0.812	20.6	11.126	282.7	2290	158	2800	193	2670	184	2800	193
12¾	323.9	110.97	165.17	0.875	22.2	11.000	279.5	2470	170	2800	193	2800	193	2800	193
12¾	323.9	118.33	176.13	0.938	23.8	10.874	276.3	2650	183	2800	193	2800	193	2800	193
12¾	323.9	125.49	186.97	1.000	25.4	10.750	273.1	2800	193	2800	193	2800	193	2800	193
12¾	323.9	132.57	197.68	1.062	27.0	10.626	269.9	2800	193	2800	193	2800	193	2800	193
12¾	323.9	139.67	208.27	1.125	28.6	10.500	266.7	2800	193	2800	193	2800	193	2800	193
*14	355.6	27.73	41.52	0.188	4.8	13.624	346.0	480	34	600	42	560	39	700	49
*14	355.6	29.91	44.93	0.203	5.2	13.594	345.2	520	36	650	45	610	42	760	53
*14	355.6	36.71	55.11	0.250	6.4	13.500	342.8	640	45	800	56	750	52	940	65
*14	355.6	41.17	61.02	0.281	7.1	13.438	341.4	720	50	900	62	840	58	1050	72
14	355.6	45.61	67.74	0.312	7.9	13.376	339.8	800	55	1000	69	940	64	1170	80
14	355.6	50.17	74.42	0.344	8.7	13.312	338.2	880	61	1110	76	1030	71	1290	88
14	355.6	54.57	81.08	0.375	9.5	13.250	336.6	960	66	1210	83	1120	77	1410	97

1		2		3		4		Test Pressure, min.							
Outside Diameter, D		Plain-End Weight, w_{pe}		Wall Thickness, t		Inside Diameter, d		5 Grade A Std		6 Grade A Alt.		7 Grade B Std		8 Grade B Alt.	
in.	mm	lb/ft	kg/m	in.	mm	in.	mm	psi	100 kPa	psi	100 kPa	psi	100 kPa	psi	100 kPa
14	355.6	63.44	94.30	0.438	11.1	13.124	333.4	1130	78	1410	97	1310	90	1640	113
14	355.6	72.09	107.39	0.500	12.7	13.000	330.2	1290	89	1610	111	1500	103	1880	129
14	355.6	80.66	120.36	0.562	14.3	12.876	327.0	1450	100	1810	125	1690	116	2110	145
14	355.6	89.28	133.19	0.625	15.9	12.750	323.8	1610	111	2010	139	1880	129	2340	162
14	355.6	97.81	145.91	0.688	17.5	12.624	320.6	1770	122	2210	153	2060	142	2580	178
14	355.6	106.13	158.49	0.750	19.1	12.500	317.4	1930	133	2410	167	2250	155	2800	193
14	355.6	114.37	170.18	0.812	20.6	12.376	314.4	2090	144	2610	180	2440	168	2800	193
14	355.6	122.65	182.52	0.875	22.2	12.250	311.2	2250	155	2800	193	2620	181	2800	193
14	355.6	130.85	194.74	0.938	23.8	12.124	308.0	2410	166	2800	193	2800	193	2800	193
14	355.6	138.84	206.83	1.000	25.4	12.000	304.8	2570	177	2800	193	2800	193	2800	193
14	355.6	146.74	218.79	1.062	27.0	11.876	301.6	2730	189	2800	193	2800	193	2800	193
14	355.6	154.69	230.63	1.125	28.6	11.750	298.4	2800	193	2800	193	2800	193	2800	193
*16	406.4	31.75	47.54	0.188	4.8	15.624	396.8	420	29	530	37	490	34	620	43
*16	406.4	34.25	51.45	0.203	5.2	15.594	396.0	460	32	570	40	530	37	670	46
*16	406.4	36.91	55.35	0.219	5.6	15.562	395.2	490	34	620	43	570	40	720	50
*16	406.4	42.05	63.13	0.250	6.4	15.500	393.6	560	39	700	49	660	46	820	57
*16	406.4	47.17	69.91	0.281	7.1	15.438	392.2	630	43	790	54	740	51	920	63
16	406.4	52.27	77.63	0.312	7.9	15.376	390.6	700	48	880	60	820	56	1020	70
16	406.4	57.52	85.32	0.344	8.7	15.312	389.0	770	53	970	66	900	62	1130	77

Table 3-1 continued

	1		2		3		4		5		6		7		8	
									Test Pressure, min.							
	Outside Diameter, D		Plain-End Weight, w_{pe}		Wall Thickness, t		Inside Diameter, d		Grade A Std		Grade A Alt.		Grade B Std		Grade B Alt.	
in.	mm	lb/ft	kg/m	in.	mm	in.	mm	psi	100 kPa	psi	100 kPa	psi	100 kPa	psi	100 kPa	
16	406.4	62.58	92.98	0.375	9.5	15.250	387.4	840	58	1050	73	980	68	1230	85
16	406.4	72.80	108.20	0.438	11.1	15.124	384.2	990	68	1230	85	1150	79	1440	99
16	406.4	82.77	123.30	0.500	12.7	15.000	381.0	1120	78	1410	97	1310	90	1640	113
16	406.4	92.66	138.27	0.562	14.3	14.876	377.8	1260	87	1580	109	1480	102	1840	127
16	406.4	102.63	153.11	0.625	15.9	14.750	374.6	1410	97	1760	121	1640	113	2050	141
16	406.4	112.51	167.83	0.688	17.5	14.624	371.4	1550	107	1940	134	1810	125	2260	156
16	406.4	122.15	182.42	0.750	19.1	14.500	368.2	1690	117	2110	146	1970	136	2460	170
16	406.4	131.71	195.98	0.812	20.6	14.376	365.2	1830	126	2280	157	2130	147	2660	183
16	406.4	141.34	210.33	0.875	22.2	14.250	362.0	1970	136	2460	170	2300	158	2800	193
16	406.4	150.89	224.55	0.938	23.8	14.124	358.8	2110	145	2640	182	2460	169	2800	193
16	406.4	160.20	238.64	1.000	25.4	14.000	355.6	2250	155	2800	193	2620	181	2800	193
16	406.4	169.43	252.61	1.062	27.0	13.876	352.4	2390	165	2800	193	2790	192	2800	193
16	406.4	178.72	266.45	1.125	28.6	13.750	349.2	2530	175	2800	193	2800	193	2800	193
16	406.4	187.93	280.17	1.188	30.2	13.624	346.0	2670	185	2800	193	2800	193	2800	193
16	406.4	196.91	293.76	1.250	31.8	13.500	342.8	2800	193	2800	193	2800	193	2800	193

Source: API Specification for line pipe, API Spec 5L, 32 Ed., March 1982

decreases. In this example for 16-in.OD pipe, the inside diameter (ID) decreases from 15.624 in. for the lightest weight pipe to 13.500 in. for line pipe weighing 196.91 lb/ft. The same relationships apply to other line pipe sizes; a variety of weights are available and wall thickness increases as weight increases.

The weight of pipe—wall thickness rather than weight is actually used in design calculations—used for a given project depends primarily on operating pressure and the design factor, or safety factor, required. For instance, when passing through a heavily populated area, a pipe with a thicker wall than would be specified in open areas for the same operating pressure may be required.

For a given operating pressure, a greater wall thickness may also be required in corrosive soil environments or when transporting corrosive fluids. For offshore pipelines, heavier pipe may be required to resist installation stresses during laying.

Service conditions also help determine what grade of pipe is to be used. Generally, the more severe the service, the higher the strength of the pipe selected. Small flow lines in oil and gas fields might typically use Grade B pipe because operating pressures are relatively low and the operating life of the pipeline is not expected to be as long as that of a gathering system or long-distance trunk or transmission line. The flow line's useful life is normally equal to the producing life of the well it serves.

Higher-strength pipe is used for pipelines operating at higher pressures, in environments that are more corrosive, and when installation stresses are high. Special grades and pipe with a particular heat-treating process or other special manufacturing step may also be required in Arctic environments. Extremely low temperatures cause some steels to be more susceptible to failure by fracture, for example.

Other types of pipe. Gathering systems for oil and gas and long-distance pipelines are all made of steel, and the individual lengths of pipe are normally joined by welding. Pipe made of materials other than steel, including fiberglass pipe, pipe made of various plastics, and cement asbestos pipe, has been used for special applications. Steel pipe with an internal lining of cement or other material has also been used.

These special pipe materials are typically used in relatively low-pressure applications and in corrosive service. Saltwater disposal pipelines are an example. Water produced with oil and separated at field-processing facilities must be disposed of in accordance with environmental and other regulations. This severe service is often better handled by materials with greater corrosion resistance than steel.

Installation of fiberglass or plastic pipe may require special procedures. Rather than being welded, individual lengths are joined by a bell-and-spigot joint, an adhesive bonded joint, a coupling that uses an O-ring to seal the joint, or other special joining methods. Because these materials are often not as ductile as steel, care may be necessary when installing them in the ditch. A sand "pad" may be needed, for instance, in rocky soil to prevent damage to the pipe. Rocks that might cause the pipe to break under the weight of backfill material often must be removed from the ditch.

These types of pipe have been available in small to medium diameters because their application involved relatively small throughputs. But at least one manufacturer now makes fiberglass pipe in sizes up to 48 in. in diameter.

Though steel pipe is used for most oil and gas field and long-distance pipelines, plastic pipe dominates gas distribution piping today.[3] In 1982, almost 80% of the more than 23,000 miles of new and replacement mains and service lines installed in the United States was plastic pipe. More than 50% of the new distribution piping installed each year since 1974 has been plastic pipe. Users of plastic pipe for these services cite several advantages: it is less expensive to buy, it is easier and more economical to install, and no cathodic protection is required to prevent corrosion.

Pipe coating

Most pipelines are coated on the exterior to protect against corrosion and other damage. The coating inhibits the flow of electric current from the pipe and the resulting loss of steel. Some pipelines are coated on the interior to improve

Fig. 3–1.   Tape is being wrapped over corrosion coating. Source: *Oil & Gas Journal,* 20 November 1978, p. 110.

flow conditions or to protect against corrosion by the fluid being transported. In addition to exterior corrosion protection coating, offshore pipelines also are coated with a layer of concrete to provide the weight needed to keep the pipeline on the sea bed.

Exterior corrosion coating. Coating and wrapping pipelines has proved to be an economical way to extend pipeline life. To be effective, a corrosion coating must resist corrosion, and it must resist damage that would uncover areas of pipe where corrosion could occur. The purpose of the wrap on the outside of the coating is to provide protection to the corrosion-inhibiting coating.

After the coating is applied, tape is wrapped around the pipe in a spiral (Fig. 3–1) with the edges overlapping slightly so all of the pipe coating is covered. Wrapping tape, normally either a heavy paper or plastic, protects the coating from damage. Coal tar enamel is the most common exterior pipe coating. Other materials include thin-film powdered epoxy, extruded polyethylene, asphalt enamel, and asphalt mastic.

An effective pipeline corrosion coating needs several properties:[4]

1. Ease of application
2. Good adhesion to pipe
3. Good resistance to impact
4. Flexibility
5. Resistance to soil stress
6. Resistance to flow (of the coating)
7. Water resistance
8. Electrical resistance
9. Chemical and physical stability
10. Resistance to soil bacteria, marine organisms, and cathodic disbondment

The use of powdered epoxy pipe coatings has grown in recent years, and they are considered to offer a number of advantages. Their resistance to chemicals, stress, and corrosion-causing action of the soil, along with improvements in the materials used in the coatings, have helped increase use of these materials.

This method has been used to coat both the interior and exterior of pipe. In one coating plant where pipe was coated both inside and out, the process included preheating, blasting and cleaning of the pipe interior and exterior, heating the pipe, applying the powder to the outside of the pipe, applying powder to the inside, quenching, and inspection (Fig. 3–2).

Epoxy powders are "thermosetting" materials; heat released by the material causes the powder to bond to the steel and form a continuous coat. Recent

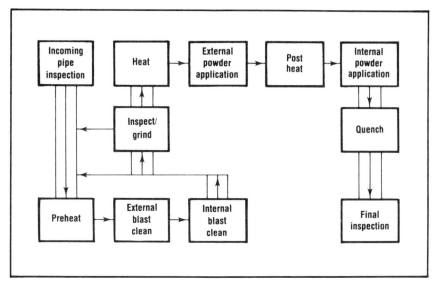

Fig. 3–2. External/internal fusion bond epoxy coating plant. Source: *Oil & Gas Journal,* 19 July 1982, p. 148.

improvements in these materials have helped widen their application. They now have better flexibility at low temperatures and greater ability to cover steel defects that were "bridged" by earlier coatings. Bridging results in areas where coating failure is likely.[5]

Pipeline corrosion coating can be applied either before the pipe is delivered to the right of way (yard applied) or after the pipe lengths are welded together and suspended above the ditch (over the ditch) as shown in Fig. 3–3. When pipe lengths are coated and wrapped at a coating yard before being delivered to the job site, a short distance at each end of each length of pipe is left bare so the joints can be welded together. When field welding is complete, coating and wrapping material is applied to the bare pipe sections.

When all coating and wrapping is done at the job site, individual lengths are first welded together and the pipeline is suspended over the ditch. Special machines then move along the pipeline on the right of way and apply coating to the entire pipe, welds and all. Tape is wrapped over the coating by a tape machine in a spiral. The wrapping machine maintains tension on the tape so it is fitted tightly over the coating.

Concrete coating. Offshore pipelines are coated with concrete in addition to the corrosion coating to provide negative buoyancy (a weight greater than the buoyant force of the water) to the pipeline. This added weight is necessary to cause the pipeline to sink to the ocean floor and remain in position on the sea

Fig. 3–3. Primer is applied to pipeline prior to coating. Source: *Oil & Gas Journal,* 15 October 1979, p. 117.

bed. To be effective, a concrete coating must resist damage during installation and after it is in place. In addition to providing needed weight, the concrete coating protects the corrosion coating.

Design of the concrete coating is critical if it is to withstand laying stresses and resist damage from anchors, fishing gear, and other hazards during operation. Considerable research has been aimed at the improvement of concrete coatings and application methods, based in part on the performance of early concrete-coated pipelines. One of the most critical considerations in concrete coating design is the overbend area where the pipe leaves the lay barge's pipe ramp during installation. If laying stresses are not properly calculated and maintained within design limits, concrete coating can crack during installation.

Other aspects of concrete coating design have also been studied. Development testing has involved the simulation of fishing trawl boards striking a concrete-coated pipeline, for instance.

Application of the concrete coating is also critical to its performance. There are several application methods available:[6]

1. *Forming,* a system more suitable for small quantities of weight coating for water crossings of short length. It is a slow process if used for large quantities of pipe.
2. *Guniting,* a process in which concrete is sprayed onto the pipe under pressure. Criteria for using this application method are similar to those for forming.

3. *Extrusion,* in which the concrete is rolled onto the pipe beneath an outer wrap. A heavy paper wrap has been used, as has a polyethylene tape. Wire reinforcement is used in this process: one layer of wire in thinner coatings and two layers of reinforcing wire in thicker coatings.

4. *Impingement,* reportedly the most widely used method of applying concrete coating. The vertically downward, high-velocity impingement application system was introduced in the early 1970s to meet requirements for a high-strength, high-impact-resistance coating. This application technique required better reinforcing, and wire mesh reinforcing was replaced with preformed cages of hard-drawn wire. The cage-type reinforcing was fitted to the corrosion coated pipe by plastic spacers, resulting in a system considered to be superior to wire mesh coatings. A typical specification for pipe diameters of 400 mm or less and for concrete coating thicknesses under 40 mm calls for cage reinforcing using 5-mm circumferential bars at 100-mm spacing with 3-mm longitudinal bars.[6]

Using the high-velocity impingement application method, concrete mixes varying in density from 140 to 190 lb/cu ft can be applied in thicknesses from 25 mm to 125 mm. The 28-day compressive strength of the coating, an important design criterion, can exceed 6,000 psi using this system.

After coating, the pipe lengths are weighed and their negative buoyancy, or submerged weight, is calculated to ensure that the submerged weight falls within tolerances specified by the user.

Concrete coatings must be designed to withstand impact and stresses even though they will be trenched after installation on the sea bed. Trenching is often not completely successful, and portions of the pipeline remain exposed. In addition, the line may be on the sea floor for some time before trenching can begin; during this time shipping and fishing activity can pose a hazard.

In one North Sea application, a test program was used as the basis for designing the concrete coating.[7] Before the test, the minimum criteria established for the coating included ability to withstand a shear of 1,200 lb/sq ft, ability to withstand 60 blows in the same place from a simulated trawl impact without damage reaching the corrosion coating beneath the concrete. Also, concrete should remain in place, even if reinforcing is exposed, for the life of the pipeline, and it should not have a significant effect on the bending behavior of the pipe.

To reduce cracking due to stresses in the overbend, a special segmented coating was designed in which the concrete was divided into 3-ft sections. Coating thickness for this project was 2 in. Tests showed that the segmented coating added very little stiffness to the pipe. Cracks during tests were only of a hairline type and closed once the test load was removed.

Fig. 3–4. Portable mill applies weight coating to 40-in. pipe. Source: *Oil & Gas Journal*, 20 November 1978, p. 110.

Concrete coating is also used in some cases to weight pipelines that cross streams (Fig. 3–4).

REFERENCES

1. Spec 5L, Specification for Line Pipe, 32 Ed., American Petroleum Institute, (March 1982); Spec 5LX, Specification for High-Test Line Pipe, 24 Ed., American Petroleum Institute, (March 1982); Spec 5LU, Specification for Ultra-High Test Heat Treated Line Pipe, (Tentative), 3 Ed., American Petroleum Institute, (March 1980); Spec 5LS, Specification for Spiral-Weld Line Pipe, 12 Ed., American Petroleum Institute, (March 1982).
2. Baldur Sommer, "Spiral-Weld Pipe Meets High-Pressure Needs," *Oil & Gas Journal*, (1 February 1982), p. 106.
3. Jim Watts, "Plastic Pipe Maintains Lion's Share of Market," *Pipeline and Gas Journal*, (December 1982), p. 19.
4. S.E. McConkey, "Fusion-Bonded Epoxy Pipe Coatings are Economic, Practical," *Oil & Gas Journal*, (19 July 1982), p. 148.
5. McConkey, See reference 4 above.
6. Eddie Kiernan, "Concrete Protects Offshore Pipelines," *Oil & Gas Journal*, (3 May 1982), p. 228.
7. J.W. Ells, "Scours and Spanning Threaten Sea Lines," *Oil & Gas Journal*, (7 July 1975), p. 67.

4

FUNDAMENTALS OF PIPELINE DESIGN

THIS discussion of the fundamentals of pipeline design is not meant to be comprehensive or to serve as a design guide. Many intricacies are involved in designing a modern pipeline system, but these are of primary interest to pipeline engineers. Rather, the goal here is to give a basic understanding of those properties of pipe and fluid and the conditions that affect the pipeline flow of gases and liquids.

The amount of fluid that must flow through the pipeline is one of the first items of information required for design. But a characteristic of many proposed pipeline projects is that future capacity requirements are difficult to forecast. Determining the capacity requirements for a pipeline gathering system to gather crude from producing fields, or natural gas from wells and processing plants, can be difficult. When oil or gas is discovered in an area, for example, several years may pass before the field is fully developed and maximum required capacity is known. Additional capacity will be needed as more wells are put on stream, but the pipeline is needed early in the field's life to transport production from the first wells.

It is also possible that more fields will be discovered within an area, significantly increasing the amount of oil or gas that must be transported by the trunk or transmission line serving the area. So additional capacity may be needed for at least a portion of the system in the future.

On the other hand, capacity requirements do not always continue to increase. After early wells are on production and a pipeline has been installed, the reservoir may not perform as well as expected. Only a few wells may be needed to develop the field; even the productivity of those wells may decline much sooner than expected. In this case, pipeline capacity requirements may decline almost from the beginning.

The same difficulties arise in forecasting for larger pipelines—crude trunk and natural gas transmission lines. But it is easier to forecast the capacity requirements when a pipeline serves a large area containing a number of fields in different stages of development and of different sizes. A production decline in one area will often be offset by an increase or a new discovery in another area, smoothing the fluctuations in capacity requirement.

In addition to difficulty in forecasting input to the pipeline, the amount of natural gas, crude, or products that must be delivered to the pipeline's customers varies seasonally and daily. Peak demand for natural gas for home heating, for instance, occurs in winter. The gas transmission pipeline must be capable of delivering these contract quantities.

Estimates of pipeline input and delivery volumes must be made, however difficult. The best projections are made of future requirements based on exploratory activity in an area, its history of oil or gas production, expected productivity of the reservoirs, and other data. Then a compromise must often be made between building a pipeline large enough to handle any possible volume requirements and one which is capable of handling only current requirements.

If too much excess capacity exists for long periods after the pipeline is built, the profitability of the system is adversely affected. If a smaller line is built and volume requirements exceed its capacity, the system must be expanded. Pipeline systems can be expanded by either adding more pumping or compression horsepower at existing stations or by building new stations, or by installing an additional pipeline along a portion or all of the route. Installing another pipeline is termed *looping*. In recent years, flow improvers have also been used to increase the capacity of large crude lines. This alternative is not yet widely used and currently has application only in crude pipelines.

Most pipelines are designed with some excess capacity or designed so capacity can be increased by the addition of compression or pumping horse-power. The increase in throughput that can be obtained by adding horsepower is limited by the maximum allowable operating pressure of the system. Maximum operating pressure is set by codes and regulations applicable to the pipe size, weight, and steel composition and by the area in which the line is located.

Pipeline design

There are several approaches to pipeline design. The most appropriate method depends on the system, the designer, the number of fixed variables, the availability of pipe and equipment, and the cost.

Both installation (capital) and operating/maintenance expense must be considered in choosing the optimum design. Often, a design having a lower installation cost than another alternative will be more expensive to operate.

When compared based on economic indicators over the life of the system, the design with the lowest installation cost may not be the best solution.

As discussed earlier, one of the most important design criteria—the volume of oil or gas to be transported—is sometimes the most difficult to determine. There is often some uncertainty in volume estimates, and making the best projection of volumes to be handled throughout the life of the pipeline is the key to a profitable project. With projected volumes and the origin and destination of the pipeline known, pipeline design typically follows these general steps:

1. A required delivery pressure is determined at the pipeline's destination. This pressure may be set by the customer's facilities or, if the line is a branch line, by the pressure required at the junction with the main line to permit fluid to flow into the main line.

2. Using the required delivery pressure and the desired throughput volume, pressure losses due to friction and the pressure required to overcome changes in elevation can be added to the delivery pressure to determine the inlet pressure. In single-phase flow, the pressure drop in the line that must be overcome by pumps or compressors is essentially the friction loss plus the pressure exerted by a liquid or gas column whose height equals the difference in elevation between the ends of the line. The pressure drop in any segment of the line is calculated in a similar manner. A trial-and-error procedure may be involved because it is necessary to choose a tentative pipe size in order to calculate pressure losses. If pressure loss is too high, the resulting inlet pressure may exceed the pressure rating of the pipe or an excessive amount of pumping or compression horsepower may be required. If this is the case, a larger pipe is selected and the calculations are repeated. The goal is to select a pipe size that can be operated efficiently at a pressure permitted by applicable regulations.

3. With the line size and operating pressure determined, the pumping or compression horsepower needed to deliver the desired volume of fluid at the specified delivery pressure can be accurately calculated. If more than one pump or compressor station is required, the location and size of additional stations is set by calculating pressure loss along the line and determining how much pump or compressor horsepower is needed to maintain operating pressure.

4. In most cases, it is necessary to perform economic calculations to compare the design with other combinations of line size, operating pressure, and horsepower in order to choose the best system.

This simplified outline represents the basic steps involved in a preliminary design of a single pipeline with no branch connections, no alternative routes, and no significant changes in throughput during its life. Few pipeline systems are that simple. Most have several branch lines feeding into a main line that consists of more than one pipe size, beginning with smaller pipe at the inlet end and requiring larger pipe as flows from the branches feed in. The volumes to be handled vary significantly over the life of the system (Fig. 4–1).

Because of this, most pipelines are now designed using sophisticated computer programs. These programs are built on basic flow equations used to design a simple pipeline manually, but the computer can perform repeated calculations on a large number of alternative solutions quickly.

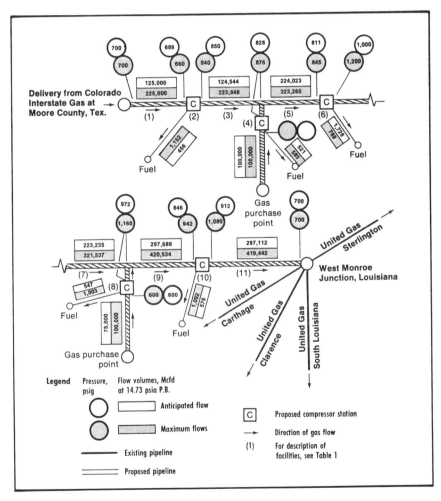

Fig. 4–1.   Example pipeline flow diagram. Source: *Oil & Gas Journal*, 14 July 1980, p. 155.

Key design terms. Even to discuss the basics of pipeline design, it is necessary to be familiar with how key physical properties of fluids affect pipeline design. It is important to remember that the term *fluid* includes both liquids and gases.

The effect of these parameters varies with the fluid; compressibility does not significantly affect the flow of liquids, for instance, and differences in viscosity among different gases may not greatly affect the flow of natural gas.

Most of the following fluid properties and other variables are considered in designing liquids or natural gas pipelines.

1. *Pipe diameter.* Of course, the larger the inside diameter of the pipeline, the more fluid can be moved through it, assuming other variables are fixed.

2. *Pipe length.* The greater the length of a segment of pipeline, the greater the total pressure drop. Pressure drop can be the same per unit of length for a given size and type of pipe, but total pressure drop increases with length.

3. *Specific gravity and density.* The density of a liquid or gas is its weight per unit volume. Density can be given in different units: In English units, it is in pounds of mass per cubic foot (lb_{mass}/cu ft); in the SI (International) metric system, units are kilograms per cubic meter (kg/cu m). The specific gravity of a liquid is the density of the liquid divided by the density of water, and the specific gravity of a gas is its density divided by the density of air. The specific gravity of air, therefore, is 1, and the specific gravity of water is 1.

4. *Compressibility.* Because most liquids are only slightly compressible, this term is usually not significant in calculating liquids pipeline capacity at normal operating conditions. In gas pipeline design, however, it is necessary to include a term in many design calculations to account for the fact that gases deviate from laws describing "ideal gas" behavior when under conditions other than standard, or base, conditions. This term, *supercompressibility factor,* is more significant at high pressures and temperatures. Near standard conditions of temperature and pressure (60°F. and 1 atmosphere, for example), the deviation from the ideal gas law is small and the effect of the supercompressibility factor on design calculations is not significant.

5. *Temperature.* Temperature affects pipeline capacity both directly and indirectly. In natural gas pipelines, the lower the operating temperature, the greater the capacity, assuming all other variables are fixed. Operating temperature also can affect other terms in equations used to calculate the capacity of both liquids and natural gas pipelines. Viscosity, for example, varies with temperature. Designing a pipeline

for heavy crude is one case in which it is necessary to know flowing temperatures accurately to calculate pipeline capacity.

6. *Viscosity.* The property of a fluid that resists flow, or *relative motion,* between adjacent parts of the fluid is viscosity. It is an important term in calculating line size and pump horsepower requirements when designing liquids pipelines.

7. *Pour point.* The lowest temperature at which an oil will pour, or flow, when cooled under specified test conditions is the pour point. Oils can be pumped below their pour point, but the design and operation of a pipeline under these conditions present special problems.

8. *Vapor pressure.* The pressure that holds a volatile liquid in equilibrium with its vapor at a given temperature is the vapor pressure. When determined for petroleum products under specific test conditions and using a prescribed procedure, it is called the Reid vapor pressure (RVP). Vapor pressure is an especially important design criterion when handling volatile petroleum products, such as LP-gas. The minimum pressure in the pipeline must be high enough to maintain these fluids in a liquid state.

9. *Reynolds number.* This dimensionless number is used to describe the type of flow exhibited by a flowing fluid. In streamlined or laminar flow, the molecules move parallel to the axis of flow; in turbulent flow, molecules move back and forth across the flow axis. Other types of flow are also possible, and the Reynolds number can be used to determine which type is likely to occur under specified conditions. In turn, the type of flow exhibited by a fluid affects pressure drop in the pipeline. In general, a Reynolds number below 1,000 describes streamlined flow; at Reynolds numbers between 1,000 and 2,000, flow is unstable. At Reynolds numbers greater than 2,000, flow is turbulent. Some references recommend, however, that flow be assumed laminar at Reynolds numbers up to 2,000 and turbulent at values above 4,000. In this case, flow is considered unstable at Reynolds numbers between 2,000 and 4,000.

10. *Friction factor.* A variety of friction factors are used in pipeline design equations. They are determined empirically and are related to the roughness of the inside pipe wall.

Other properties of the fluid and pipe may be used in specific calculations, but these are the basic terms used to determine pressure drop and flow capacity. Many system variables are interdependent. For example, operating pressure depends, in part, on pressure drop in the line. Pressure drop, in turn, depends on flow rate, and maximum flow rate is dictated by allowable pressure drop.

Several pressure terms are used in pipeline design and operation. Barometric pressure is the value of the atmospheric pressure above a perfect vacuum. A perfect vacuum cannot exist on the earth, but it makes a convenient reference point for pressure measurement.

Absolute pressure is the pressure of a pipeline or vessel above a perfect vacuum and is abbreviated *psia*. Gauge pressure is the pressure measured in a pipeline or vessel above atmospheric pressure and is abbreviated *psig*. Standard atmospheric pressure is usually considered to be 14.696 lb/sq in., or 760 mm of mercury, but atmospheric pressure varies with elevation above sea level. Many contracts for the purchase or sale of natural gas, for instance, specify that standard, or base, pressure will be other than 14.696 lb/sq in.

Formulas describing the flow of fluids in a pipe are derived from Bernoulli's theorem and are modified to account for losses due to friction. Bernoulli's theorem expresses the application of the law of conservation of energy to the flow of fluids in a conduit.[1] To describe the actual flow of gases and liquids properly, however, solutions of equations based on Bernoulli's theorem require the use of coefficients that must be determined experimentally.

The theoretical equation for fluid flow neglects friction and assumes no energy is added to the systems by pumps or compressors. Of course, in the design and operation of a pipeline, friction losses are very important, and pumps and compressors are required to overcome those losses (Fig. 4–2). So practical pipeline design equations depend on empirical coefficients that have been determined during years of research and testing.

The basic theory of fluid flow does not change. But modifications continue to be made in coefficients as more information is available, and the application

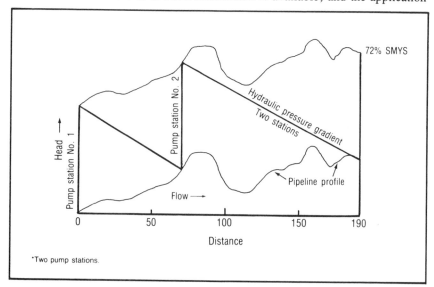

Fig. 4–2.   Pipeline profile and hydraulic gradient. Source: *Oil & Gas Journal,* 15 June 1981, p. 132.

of various forms of basic formulas continues to be refined. The use of computers for solving pipeline design problems has also enhanced the accuracy and flexibility possible in pipeline design.

Liquids pipelines

In the design of both liquids and natural gas pipelines, pressure drop, flow capacity, and pumping or compression horsepower required are key calculations. The design of a liquids pipeline is similar in concept to the design of a natural gas pipeline. In both cases, a delivery pressure and the volume the pipeline must handle are known. The allowable working pressure of the pipe can be determined using the pipe size and type and specified safety factors.

In most pipeline calculations, assumptions must be made initially. For instance, a line size may be assumed in order to determine maximum operating pressure and the pressure drop in a given length of pipe for a given flow volume. If the resulting pressure drop, when added to the known delivery pressure, exceeds the allowable working pressure, a larger pipe size must usually be chosen. It may be possible to change the capacity and spacing of booster pumping stations to stay within operating pressure limits. But in the simplest case, if the calculation yields an operating pressure greater than allowed, a larger pipe size must be selected and the calculation repeated.

It is apparent that many options are available in even a moderately complex pipeline system. But today's computer programs for pipeline design can analyze many variables and many options in a short time, greatly easing the design process.

Pressure drop. As discussed earlier, Bernoulli's theorem describes the flow of fluids—gases and liquids—in a pipe. The general equation for the flow of liquids in a pipe is Darcy's formula. To determine pressure drop, for instance, the equation is used in this form:

$$\Delta P = \frac{\rho f L v^2}{144 \, D \, 2g}$$

Where:

ΔP = pressure drop over length L, psig
ρ = density of fluid, lb/cu ft
f = friction factor, dimensionless
L = length of pipe, ft
v = velocity of flow, ft/sec
D = inside diameter of pipe, ft
g = acceleration of gravity = 32 ft/sec^2

In this equation, L can be either the entire length of the pipeline or another specified length. If the length of the entire pipeline is used, ΔP is the total pressure drop in the line. If, however, L is chosen as, say, 100 ft, ΔP is the pressure drop in psig/100 ft.

Since flow velocity is a function of flow volume (in gal/min or similar units), velocity can be determined using the desired flow volume and an assumed pipe size. Then that assumed pipe size and calculated velocity are used in the equation to determine pressure drop.

If a pressure drop is assumed, using the required delivery pressure and flow volume, the line size required can be calculated. Again, the assumed pressure drop must be checked using the calculated line size to ensure allowable operating pressure is not exceeded.

The Darcy equation can be derived mathematically except for the friction factor, f. This factor must be determined experimentally.

The Darcy equation can be used for both laminar and turbulent flow of liquid in a pipe. But if the pressure in the pipe falls below the vapor pressure of the liquid, such as might be the case with light hydrocarbon liquids, flow rates calculated using the equation will not be accurate. Cavitation occurs when pressure falls below the liquid's vapor pressure.

Empirical equations are also used to describe the flow of liquids in a pipe. The Hazen and Williams equation for the flow of water, for example, is:

$$Q = 0.442 \ d^{2.63} \ c \left(\frac{P_1 - P_2}{L} \right)^{0.54}$$

Where:

- Q = flow volume, gal/min
- d = inside diameter of pipe, in.
- P_1 = inlet pressure, psig
- P_2 = outlet pressure, psig
- L = pipeline length, ft
- c = a constant depending on pipe roughness ($c = 140$ for new steel pipe)

Valves and fittings. In addition to the pressure loss due to friction of the flowing fluid with the walls of the pipeline, valves and fittings also contribute to overall system pressure loss. The pressure loss due to a single valve in several thousand feet of straight piping will be relatively insignificant. But in a pumping station, for example, where many valves exist and many changes in flow direction occur, pressure loss in valves and fittings is important.

Pressure loss in valves and fittings is made up of both the friction loss within the valve or fitting itself and the additional loss upstream and downstream of the fitting above that which would have occurred in the absence of the fitting. Calculation of the pressure loss in a valve or fitting is based on experimental data. One approach is the use of a resistance factor for a given valve or fitting.[2] The resistance coefficient is normally treated as a constant for a given valve or fitting under all flow conditions.

Another term used in determining pressure drop through valves and fittings is the flow coefficient, C_v. The flow coefficient of a valve is the flow of water at 60°F., in gal/min, at a pressure drop of one psi across the valve. The flow coefficients of any other liquid can be calculated using the relation of its density to that of water.

Heavy crude.   The type of crude must be considered in pipeline design because viscosity and other physical properties affect throughput and pumping calculations. For most crudes, no special equipment is required in the pipeline system for different types of crudes. But there are some crudes with very high pour points or high wax contents that require pipelines of special design.

Pour point can indicate the amount of different types of hydrocarbons in the crude.

Pipelines handling these crudes are usually short, typically connecting a well to a production platform offshore or to crude-treating facilities in onshore fields. Pipelining such crudes can be especially troublesome offshore where heat loss to the water is great. Heat added to the crude before it enters the pipeline is dissipated within a short distance if a conventional pipeline is used. If the crude cools, excessive wax deposits in the pipeline can lower operating efficiency. In cases of extremely viscous crudes, flow can even be halted if the temperature is allowed to fall too low. Not only is the halting of flow a problem, but restarting flow after such an occurrence can be difficult.

To handle these special crudes, pipelines have been successfully installed and operated. A recent example of such a project involved three thermally insulated pipelines laid in about 115 ft of water offshore Gabon.[3] Two 1.24-mile-long lines connected drilling/production platforms and a 2-mile line connected a production platform with a storage tanker moored in the field. The three pipelines were each 10¾-in. diameter with rigid polyurethane foam insulation contained in a high-density polyethylene sleeve. An insulated pipeline is not the only solution to transporting heavy, high-pour-point oil. Other approaches include the following:[4]

1. Heating the crude to a high temperature at the inlet to the pipeline, allowing it to reach its destination before cooling below the pour point. The pipeline may or may not be insulated.
2. Pumping the crude at a temperature below the pour point.
3. Adding a hydrocarbon diluent such as a less waxy crude or a light distillate.
4. Injecting water to form a layer between the pipe wall and the crude.

5. Mixing water with the crude to form an emulsion.
6. Processing the crude before pipelining to change the wax crystal structure and reduce pour point and viscosity.
7. Heating both crude and pipeline by steam tracing or electrical heating.
8. Injecting paraffin inhibitors.

A combination of these methods can also be used to transport heavy oils by pipeline. The choice of method to use involves consideration of the physical properties of the crude, heat transfer, restart after shutdown, and facilities design.

Waxy crude can be pumped below its pour point; more pumping energy is required, but there is no sudden change in fluid characteristics at the pour point as far as pumping requirements are concerned. If pumping is stopped, more energy will be required to put the crude in motion again than was required to keep it flowing. When flow is stopped, wax crystals form, causing the crude to gel in the pipeline. However, the additional energy required to restart flow will be less than if the crude had been pumped above the pour point and allowed to cool down after flow had stopped.[4] Experiments have shown that restart pressures can be 5–10 times higher for a pipeline that was above the pour point and cooled after shutdown than for one that was below its pour point before shutdown.

Gas pipelines

Several formulas can be used to calculate the flow of gas in a pipeline. These formulas account for the effects of pressure, temperature, pipe diameter, pipe length, specific gravity, pipe roughness, and gas deviation.

The Darcy equation can also be used in flow calculations involving gases, but it must be done with care and restrictions on its use are recommended. If, for instance, pressure drop in the line is large relative to the inlet pressure, the Darcy equation is not recommended. Because this is often the case and because other restrictions also apply to its use in gas flow calculations, other more practical equations are commonly used for gas flow calculations.

Gas mixtures.   An early step in gas pipeline or gas compressor design is an analysis of the gas stream to be transported or compressed. As it comes from the gas well and is transported through gathering and transmission pipelines, natural gas is a mixture of several components. At different points in a gas pipeline system, the amount of each component changes. For instance, in the field flow line, the gas stream may contain large amounts of gas liquids, the "heavier" components. But large volumes of these heavier components may be removed in the gas-processing plant, so gas in the gas transmission line is of a much different composition.

Methane makes up the largest share of most natural gas streams, but significant amounts of ethane and propane may also be present, as well as lesser amounts of butanes, pentanes, and heavier components. Also present in some natural gas streams are nitrogen, carbon dioxide, hydrogen sulfide, and water.

Each of these components has different physical properties. Use of the physical properties of a single component in calculations involving the mixture would give inaccurate results. It is therefore necessary to calculate the physical properties of the mixture before performing flow or other calculations. Specific gravity of the mixture, for example, is needed in gas flow equations. Other properties are required for other design steps, including ratio of specific heats, pseudocritical temperature and pressure, and heating value.

The amount of each component in the gas stream can be determined using several instruments, including a mass spectrometer, infrared analyzers, or a gas chromatograph. In the gas chromatograph, the most common method, a sample of the gas is passed through a column with a carrier gas, typically helium or air. Different components of the gas mixture exit the column at characteristic intervals, and detectors in the carrier gas stream record the quantity of each component.

Information recorded by the chromatograph can provide the composition of the gas mixture by expressing the amount of each component as a fraction or percentage of the mixture. The sum of the fractions of each component equals 1.

Finding the specific gravity of a mixture, a parameter needed in gas flow calculations, is an example of how the properties of a gas mixture can be determined from gas analysis data. The specific gravity of the mixture is the ratio of the molecular weight of the mixture to the molecular weight of air (air = 28.964 lb_{mass}/lb-mole). The contribution of each component to the specific gravity of the mixture is first found by multiplying each component's molecular weight by the fraction of that component in the mixture. The sum of these individual contributions is then divided by the molecular weight of air to obtain the specific gravity of the gas mixture. Columns 1, 2, and 3 in Table 4–1 show this calculation.

Another way to determine the specific gravity of a gas mixture is to use the specific gravity of the individual components. The specific gravity of a component is multiplied by the fraction of the component in the gas mixture to determine the contribution of each component to the specific gravity of the mixture. Then these contributions are totaled to give the mixture's specific gravity. Columns 1, 4, and 5 in Table 4–1 illustrate this procedure.

Other properties of the gas mixture can be determined in a similar way by multiplying the desired physical property of the individual component by its fraction in the gas stream, then summing those multiplication products.

Allowable operating pressure.   An important pipeline design calculation is the maximum pressure at which a given size, grade, and weight of pipe may

TABLE 4-1
Calculating Specific Gravity Of A Gas Stream

Component	Molecular weight (from table of physical properties)	Molecular fraction (from gas analysis)	Share of molecular weight of mixture (1×2)	Specific gravity of component (from table of physical properties)	Share of specific gravity of mixture* (2×4)
Methane (CH_4)	16.043	0.692	11.102	0.554	0.383
Ethane (C_2H_6)	30.070	0.153	4.601	1.038	0.159
Propane (C_3H_8)	44.097	0.066	2.910	1.523	0.101
Isobutane (iC_4H_{10})	58.124	0.006	0.349	2.007	0.012
Normal butane (nC_4H_{10})	58.124	0.014	0.814	2.007	0.028
Isopentane (iC_5H_{12})	72.151	0.005	0.361	2.491	0.013
Normal pentane (nC_5H_{12})	72.151	0.003	0.217	2.491	0.008
Hexane (C_6H_{14})	86.178	0.004	0.345	2.975	0.012
Heptane (C_7H_{16})	100.205	0.002	0.200	3.460	0.007
Carbon dioxide (CO_2)	44.010	0.020	0.880	1.520	0.030
Nitrogen (N_2)	28.013	0.035	0.981	0.967	0.034
		1.000	22.760		0.787

*Specific gravity mixture $= \dfrac{\text{Molecular weight mixture}}{\text{Molecular weight air}} = \dfrac{22.760}{28.964} = 0.786$

operate. Maximum operating pressure determines how much gas a pipeline may carry, other factors being fixed, and depends on the physical and chemical properties of the pipe steel. Since standard pipe grades, sizes, and weights are normally used, the maximum operating pressure can usually be obtained from tables contained in recognized specifications.

But these design pressures have been determined using the basic formula for calculating *hoop stress* in a cylinder (pipe). The allowable operating pressure for pipe sizes or grades not contained in tables can be calculated using this formula. For natural gas pipelines in the United States, for example, the equation includes design factors to account for the presence of longitudinal welds in the pipe; the location of the pipeline relative to populated areas; and excessively high temperatures. These factors lower the allowable operating pressure, providing a safety margin. With these factors incorporated into the basic hoop stress formula, pipeline design pressure is determined by:

$$P = \frac{2St}{D} \times F \times E \times T$$

Where:

P = design pressure, psig

S = minimum yield strength stipulated in the specifications under which the pipe was manufactured

D = nominal outside diameter of the pipe, in.

t = nominal wall thickness of the pipe, in.

F = construction type design factor

E = longitudinal joint factor

T = temperature derating factor

This equation can also be used to calculate the required pipe wall thickness for a given operating pressure and pipe size.

Construction factors (F) used in this equation are designated in USA Standard B31.8, Code for Pressure Piping, Gas Transmission and Distribution Piping Systems for four areas:

1. Type A construction is that done in Class 1 locations, which include wastelands, deserts, rugged mountains, farm land, and similar areas. The factor, F, for designing pipelines for these areas is 0.72.

2. Type B construction, Class 2 locations, includes fringe areas around cities or towns, and farm or industrial areas with a specified population density. This class is between Class 1 and Class 3 locations, and the construction type factor is 0.60.

3. Type C construction, Class 3 locations, includes areas subdivided for residential or commercial purposes with a specified building density of a specified type. The construction type factor is 0.50.

4. Type D construction, Class 4 locations, includes areas where multistory buildings are prevalent, where traffic is heavy or dense, or where there are numerous other underground utilities. The Type D construction factor is 0.40.

Construction type factors are also specified in pipeline codes for pipe that crosses or runs parallel to roads and railroads.

The effect of the construction type factor is to lower the allowable operating pressure for a given size, weight, and grade of pipe as the construction area becomes more populated and the failure of a pipeline would be more serious. The other two factors in the design pressure formula also derate the allowable operating pressure, providing an additional safety margin. The longitudinal joint factor varies with the type of joint used in manufacturing the pipe; for seamless pipe and some longitudinally welded pipe, the factor is 1. There is no longitudinal joint in seamless pipe; it is essentially a hoop. A factor of 1 permitted when using some longitudinally welded pipe means the weld is as strong as a seamless pipe. When pipe is manufactured by other longitudinal welding methods, however, a factor of 0.60 or 0.80 must be used to calculate maximum allowable operating pressure.

The temperature derating factor, T, for steel pipe varies from 1 for operating temperatures of 250°F or less to 0.867 for an operating temperature of 450°F.

There are other specifications and limits that may apply to pipeline pressures. Limits on operating pressure may be set by the fluid being transported. For example, LP-gas and natural gas liquids pipelines must be operated at pressures above the vapor pressure of the fluid. Operating pressure is also an important criterion in CO_2 pipeline design. In all cases, pressures specified in applicable codes and regulations must not be exceeded.

Friction losses.   A friction factor, one of the empirical terms in flow equations, is used to determine the pressure loss due to friction in natural gas pipelines just as is the case in liquids pipelines. Internal losses, which depend on the viscosity and gravity of the gas, the flow rate, and the pipe diameter, must also be considered. These internal losses within the flowing fluid vary according to the type of flow, or *flow regime*, that exists in the pipeline. The Reynolds number helps determine which flow regime exists. To calculate the Reynolds number for a natural gas pipeline, this equation may be used:[5]

$$Re \simeq \frac{20\ QG}{\mu D}$$

Where:
Q = gas flow rate at 60°F and 14.73 psia, Mcfd
G = gas gravity (air = 1)
μ = gas viscosity at flow temperature and pressure, cp
D = pipe diameter, in.

Pressure losses that occur due to friction with the pipe depend on the roughness of the pipe wall. The absolute roughness, e, is the distance from peaks to valleys on the inside surface of the pipe and is given in inches. Line pipe, for instance, typically has a roughness of 0.0007 in. By comparison, e for the rougher cement-lined pipe is 0.01 to 0.1 in. The relative roughness is the ratio of this distance to the pipe diameter, or e/D.

Calculating gas flow. Several equations are accepted for calculating flow capacity of gas pipelines. The Weymouth equation is considered by some designers to be most applicable to smaller-diameter lines (15 in. or less). The Panhandle equation and Modified Panhandle equation are considered appropriate for larger gas pipelines.[5]

A formula developed by the American Gas Association (AGA) is considered by some to be superior now that computer programs are available that can routinely solve this more complex equation. The AGA formula involves the calculation of a transmission factor based on the flow regime and other parameters and takes into account changes in elevation.[6] The first three equations discussed are as follows:[7]

Weymouth

$$Q = 433.5 \, \frac{T_b}{P_b} \left[\frac{P_1^{\,2} - P_2^{\,2}}{GTLZ} \right]^{0.5} D^{2.667}$$

Panhandle

$$Q = 435.7 \left(\frac{T_b}{P_b} \right)^{1.0788} \left[\frac{P_1^{\,2} - P_2^{\,2}}{G^{0.8539} TLZ} \right]^{0.5392} D^{2.6182}$$

Modified Panhandle

$$Q = 737 \left(\frac{T_b}{P_b} \right)^{1.02} \left[\frac{P_1^{\,2} - P_2^{\,2}}{G^{0.961} TLZ} \right]^{0.51} D^{2.53}$$

Where:

Q = flow rate (units vary, depending on the units used for other terms in the formula; constants will also differ when different units are used)

P_b = base pressure, psia

T_b = base temperature, °F absolute (°R)

P_1 = inlet pressure, psia

P_2 = outlet pressure, psia

G = gas specific gravity (air = 1)
T = average flowing gas temperature, °F absolute (°R)
L = length of pipe (feet or miles, depending on formula and units used)
D = inside diameter of pipe, in.
Z = gas supercompressibility factor (gas deviation factor)

The main difference in these formulas is the size range in which they have been found to be most applicable and the treatment of pipe friction. In the basic flow equation for isothermal flow in a horizontal line, a friction factor term is included. In developing the above equations for flow of natural gas, this friction factor has been incorporated by relating it to other flow parameters. The friction factor used when this is done accounts, in part, for the variety of constants and exponents among the equations.

A number of investigations have developed empirical friction factors for various conditions. Choice of the proper factor—indeed, choice of the proper flow equation for specific conditions—requires expertise in pipeline design.

Most of the terms described above have been explained previously. One further explanation is appropriate: °R is degrees Rankine and represents a different temperature scale. The temperature measurement is common in gas flow equations and is determined by adding 459.6 to the temperature in degrees Fahrenheit.

The AGA equation for calculating pipeline flow is somewhat more complex than the three equations discussed earlier but involves the same basic parameters:[6]

$$Q = 38.77 \left(\frac{T_b}{P_b} \right) F \left[\frac{P_1^2 - P_2^2 - \left(\frac{0.0375 \ GHP_m^2}{Z_m \ T_f} \right)}{GT_f Z_m L} \right]^{0.5} D^{2.5}$$

Where:
F = a transmission factor that is based on the flow regime and other variables, $F = 4 \log_{10} (3.7D/K_e)$
K_e = the relative roughness of the pipe
P_m = the mean pressure in the pipeline
Z_m = the supercompressibility factor at that mean pressure
H = the change in elevation of the pipeline between inlet and outlet

In gas computations, base temperature and pressure are the temperature and pressure at which a cubic foot (or cubic meter) is the unit of measurement. These base conditions are necessary to provide a standard for gas measurement because the volume of a gas varies with its temperature and pressure. Base temperature and pressure are usually stipulated in gas purchase contracts and other documents. Typical standard conditions include a base pressure of 14.73 psia and a base temperature of 60°F.

As mentioned earlier, the calculation of pressure losses and flow rates depends on pipeline size; pipeline size required, in turn, depends on pressure loss and volume. So it is often necessary to make a preliminary choice of pipeline size before detailed calculations are made on flow rate and pressure drop. Charts are available that give a rough estimate of the volume of gas that can flow through a given pipe size and weight with a given pressure drop (typically expressed in psi/100 ft).

Knowing the flow rate required and assuming a reasonable pressure drop based on experience, the experienced designer can choose a likely pipe size as a starting point. Then, calculation of pressure drop and flow capacity can be made assuming that size and weight of pipe. After these calculations are made, a change in the pipe size may be necessary to adjust operating pressure, or a change in pipe wall thickness may be needed to meet requirements dictated by operating pressure.

With computer programs in common use today, these choices can be evaluated rapidly and the correct design selected.

A system.   The procedures described briefly here apply to the design of a single line or section of line that has no branches. But most pipeline design problems involve a number of branches, sections of different pipe diameters and weights, and other complexities. Compression and pumping must also be considered. Few pipelines operate without compression or pumping, and the combination of pipeline and pumps or compressors must be designed as a system.

Each is an integral part of the optimum design. Pipeline design will affect the size and number of compressors or pumps required; compressor or pump-station design will affect pipeline operating conditions (see Chapter 5).

Two-phase pipeline design

The design of a two-phase pipeline to handle both gas and liquids involves calculations similar to those used for a single-phase pipeline. The goal in both cases is to determine pipe size, flow capacity, pressure drop, and other flow parameters.

The key difference is that pressure drop is much more difficult to determine when both gas and liquid are flowing in the same pipeline. Flow of the two phases can take several forms, and pressure drop can vary widely, depending on flow conditions. Changes in elevation over the route of a two-phase line are much more significant than in a single-phase pipeline.

Besides pressure drop, liquid holdup is an important consideration in the design of a two-phase pipeline. *Holdup* refers to the fraction of the pipeline occupied by liquid at any point in the line and is a function of liquid and gas flow rates, fluid properties, pipeline slope, and time.[8]

Flow regime (Fig. 4–3) is a term used to characterize how the liquid and gas flow within the pipeline. In bubble flow, free gas is present as bubbles in a continuous liquid phase. At the other extreme is mist flow, in which the gas phase is continuous and liquid droplets are entrained in the gas.[9] Between these two extremes are other types of flow, including *stratified wavy* and *slug flow*. In slug flow, at low flow rates, liquid can occupy the entire cross section of the pipeline at points in the line. This is likely to occur on uphill portions of the pipeline. This type of flow can produce liquid slugs that exit the pipeline intermittently. Because of this, it is often necessary to include equipment to catch these slugs of liquid at the end of the pipeline to prevent damage to processing or other facilities.

Much of the data used to design two-phase pipelines have been determined experimentally and through tests made in operating two-phase pipelines. Two-phase pipeline design is a subject on which research and testing continue, and

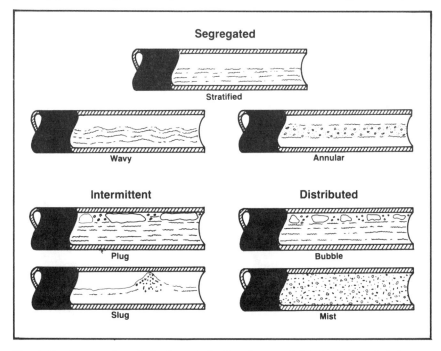

Fig. 4–3. Flow regimes in horizontal pipeline. Source: *Oil & Gas Journal*, 9 August 1982, p. 132.

sophisticated computer programs have been developed to predict flow conditions and pressure drop more accurately.

Two-phase pipelines have been built and operated successfully, though the simpler approach is to use two single-phase pipelines: one transporting liquids and the other gas. There are applications, however, in which the construction of two pipelines along the same route is the least economic solution. The most common application of two-phase pipelines is offshore, where pipeline construction costs are high. In this application, the two-phase pipeline—even though it is more difficult to design and operate and may be more costly to install than one single-phase pipeline—is the most economical approach.

Arctic pipeline design

Several special considerations must be included in designing an oil or gas pipeline for low-temperature areas. Low-temperature steels may be required, for example. The presence of permafrost areas may dictate whether or not the line is buried, the need for cooling, and other special design and operating techniques (see Chapter 7).

The operating temperature of an Arctic pipeline is a critical design parameter. Temperatures must be kept below freezing to avoid permafrost melting, which can cause the pipeline to be stressed beyond design limits. In other areas where soil is not frozen, pipelining a cold fluid can cause *frost heave*, again putting undue stress on the pipeline.

The trans-Alaska crude pipeline was installed aboveground on special supports through the permafrost area so the hot oil would not melt the permafrost. The line was buried in areas where permafrost does not exist.

The proposed Alaska Natural Gas Transportation system was designed to be cooled to below the freezing point to protect permafrost and increase flow efficiency. Operating temperature was set between a maximum of 28°F and a minimum of 0°F. This operating range prevents thawing of the permafrost but is within the fracture control limits specified for the pipe steel and avoids condensation of liquids in the line. Refrigeration compressors at the gas compressor stations along the pipeline would provide cooling to maintain the gas at the desired temperature.

Pipelining in cold climates also poses the danger of the formation of hydrates in natural gas pipelines and the crystallization of wax in crude lines. The friction due to a buildup of hydrates or wax crystals can increase pumping requirements and decrease efficiency. Viscosity of the crude being pumped, a key factor in designing Arctic pipelines, is a function of temperature and an important criterion in calculating the pumping energy required.

In crude lines, consideration must also be given to the possibility of a shutdown. Special techniques or equipment may be needed to start up a pipeline after it has been shutdown because the crude—depending on temperature, viscosity, etc.—will tend to congeal in the line. On portions of the trans-Alaska crude pipeline, insulation designed to retain enough heat to start up the line after a 21-day shutdown was required.[10] To start up a line that has become too viscous during a shutdown, chemicals can be added at the inlet to inhibit wax crystal formation, the line can be heat-traced to melt the wax, or high pump pressures can be used to start flow. High pressures are often not a practical solution because the pipe rating and pump capacity may be exceeded by the pressure required to start the line.

Preventing the formation of hydrates in natural gas lines can be done by injecting methanol into the pipeline.

Other areas also need special attention in designing a pipeline for severe environments. Insulation must meet the specified heat loss limits and often must be watertight. Heavy snow loads must be included in design calculations for stations and other facilities in Arctic areas. And in all phases of Arctic pipeline design, logistics and the working environment must be considered. The fact that personnel will be working in heavy clothing means certain operations performed in more moderate climates are impractical. Productivity in Arctic areas is necessarily lower than in other environments, and special considerations are also needed in scheduling construction work.

The special design problems discussed here for cold-weather environments are, of course, required in addition to the basic calculations of flow capacity, pressure drop, and pipe size.

Energy efficiency

The rapid increase in oil and gas prices during the 1970s brought emphasis on reducing the use of energy in all petroleum operations, including pipeline systems. Not only has efficiency become an important part of the design of a new system, but today's oil and gas prices make revamping of existing systems or replacing old equipment with more energy-efficient equipment economically justifiable. Payout for replacement equipment can be quick in many cases.

In general, many decisions on designing for energy efficiency or for increasing the efficiency of existing systems hinge on the tradeoff between larger line sizes and greater pumping or compression horsepower. Engine fuel is typically the largest use of energy in a pipeline system. In the case of natural gas pipeline systems, for example, studies have shown that beyond a certain operating pressure the use of larger pipe—or *looping* portions of existing pipelines—to increase system capacity is less expensive than additional compression horsepower and saves significant amounts of fuel.[11] (A loop is a pipeline laid along the same general route as an existing pipeline to increase the capacity of the system.)

Even though energy consumption in a pipeline system is a very small fraction of the energy contained in the oil or gas transported by the pipeline, that amount can be significant in large systems. For instance, Alberta Gas Trunk Line of Canada (now NOVA) estimated in 1979 that it consumed only about 1% of the gas transported by its system, but that amounted to about 16.6 billion cu ft annually.

In the late 1970s, NOVA was using a number of approaches to reduce energy consumption in its system. Though other techniques can be used, these represent key considerations that can be a part of the design of new systems and the study of the feasibility of revamping existing natural gas pipeline systems.

In 1977, NOVA added about 23 miles of loop to its system, saving about 2.4 billion cu ft of gas per year. Internal coating of pipelines can reduce pressure drop that has to be made up by compressor horsepower by reducing internal pipe roughness. NOVA uses internal epoxy coating on much of its system and in 1979 estimated the practice saves about 900 million cu ft/year of gas.

Since compressors on the system are the largest fuel consumer, the most efficient units are installed when new capacity is required. Although capital and maintenance costs are higher in some cases, lower fuel consumption makes the more efficient units economical. The gas saved quickly pays for the cost differential. NOVA also uses mobile compressors to fit changing volume requirements more closely.

Another use of compressors to save fuel is the recovery of gas that would otherwise be vented to the atmosphere when additions to the system are tied in or repair work is required. By using this portable "pulldown" compressor, gas in the section to be isolated can be recovered and pumped into another line or routed back into the line downstream of the isolation valve.

Optimizing the operation of a pipeline system through the use of computer monitoring and control is also an effective way to save energy. It can help schedule and coordinate maintenance and reduce downtime.

NOVA also cites several approaches to compressor station design that can save energy:[12]

1. Use proper pipe diameters and full-ported valves to keep piping pressure losses to a minimum.
2. Design scrubbers and other vessels in the station for a minimum pressure loss.
3. Use measurement devices other than orifice meters to reduce compressor station pressure losses further.

Other design problems.   Determining pipe size, pressure drop, flow capacity, and related criteria are only the basics of pipeline design. Many other aspects of a pipeline system require sophisticated, detailed design work, including the following:

1. Steel selection is critical in special environments such as the Arctic and offshore. Design can involve the consideration of special tests of pipeline steels to determine toughness, susceptibility to fracture and fracture propagation, and other steel characteristics.
2. Calculation of laying stresses is an important part of offshore pipeline design. Pipe stresses are often greater during installation than at any other time. Insuring that irregularities on the ocean floor do not overstress the pipeline after it is laid is also important, requiring consideration of the type of lay barge, water depth, pipe weight, and other factors.
3. Calculating buckling and collapse resistance of an offshore pipeline— both during installation and operation—is a key step in the design process.
4. Many pipelines undergo significant temperature changes during operation that cause expansion of the pipe. Some pipelines carry warm or hot fluids, and when shut down, cooling occurs. Allowing for expansion and contraction of the steel can be done in several ways.
5. Calculating pump and compressor horsepower required is an integral part of pipeline flow and pressure drop calculations when designing a complete system. Pump and compressor calculations involve a unique group of equations and special equipment (see Chapter 5).
6. Cathodic protection, widely used for corrosion control in pipelines, is a key part of the design of many systems. It helps ensure the pipeline will be serviceable throughout its intended life.
7. Stream and road crossings may require sophisticated design. For example, large pipeline bridges required for spanning rivers can present complex structural design problems.
8. Control system design is a science in itself. It involves monitoring the pipeline for leaks, and monitoring and controlling pumps and other operating conditions.
9. Station design, in addition to the sizing of pumps or compressors, can involve the design of buildings and other structures.
10. Economic calculations are a key part of pipeline design. An estimate of capital costs and operating costs must be made, and alternative designs compared.

Other special problems are not uncommon in pipeline design, but those outlined here give an idea of the complexity of the process for a modern pipeline system.

REFERENCES

1. "Flow of Fluids Through Valves, Fittings, and Pipe," Technical Paper No. 410, Crane Co., 1979.
2. See reference 1 above.
3. Mike Hales, "Thermally Insulated Pipelines Successfully Move High-Wax-Content Crude Offshore Gabon," *Oil & Gas Journal,* (25 January 1982), p. 179.
4. Bill Smith "Pumping Heavy Crudes—1: Guidelines Set Out For Pumping Heavy Crudes," *Oil & Gas Journal,* (28 May 1979), p. 111.
5. Chi U. Ikoku, *Natural Gas Engineering: A Systems Approach,* Tulsa: *Penn-Well Publishing Co.,* 1980.
6. Jack N. Rau and Mike Hein, "AGA Gas Flow Equations Can Be Quickly Done with Program for Hand-Held Calculators," *Oil & Gas Journal,* (8 March 1982), p. 233.
7. *Gas Engineers Handbook,* 1 Ed., New York: Industrial Press, 1974.
8. E.V. Seymour, "North West Shelf Pipeline Design—1: Design Detailed for Australia's North West Shelf Pipeline," *Oil & Gas Journal,* (31 August 1981), p. 51.
9. See reference 8 above.
10. G.L. Carlson and G.C. Cheap, "Innovations Met Trans-Alaska Line Insulation Requirements," *Oil & Gas Journal,* (27 February 1978), p. 128.
11. Ralph C. Hesje, "AGTL Program Boosts Energy Efficiency in Gas Transmission," *Oil & Gas Journal,* (23 July 1979), p. 43.
12. See reference 11 above.

5

PUMPS AND COMPRESSORS

PUMP stations for liquids pipelines and compressor stations for natural gas
pipelines are just as important a part of the pipeline system as the pipeline
itself.

There are two general types of pump or compressor stations in a pipeline
system. The originating station at the inlet to the pipeline is usually the most
complex; booster stations along the pipeline contain less equipment. Each type
of station contains either liquid pumps in crude or products pipeline systems, or
gas compressors in natural gas pipeline systems.

Station design and operation

In designing a pipeline system, the location of pump or compressor stations
must be determined as well as the size of individual pumps or compressors
within each station. The number and location of stations depends on the length
of the pipeline and how much energy must be added to the fluid to transport the
required volume at the desired delivery pressure.

In general, fluid leaves a pump or compressor station at a given discharge
pressure. As the distance from the station increases, the pressure in the pipeline
decreases due to friction and elevation losses. Pressure in the pipeline cannot be
allowed to fall below the delivery pressure at the end of the pipeline or flow will
cease. So if the length of the line is such that pressure drop will consume enough
of the energy supplied by the pump or compressor to drop operating pressure
below the delivery presssure, another pump or compressor station—a booster
station—is required.

If the pipeline is a branch line feeding into a main line, pressure in the branch must be maintained high enough to cause fluid to flow into the main line.

The number of booster stations varies widely in both natural gas and liquids pipeline systems. The longer the line, the more stations may be required. The size of these stations—the amount of pump or compressor horsepower contained in each—also varies widely. Most large systems with a number of pump or compressor stations represent a compromise between a few very large stations and a large number of small booster stations. The advantage of a few stations is that operation, control, and maintenance are more centralized. But if, for instance, a single, very large booster station were used, it might have to increase the pressure at the station to a level that would require the use of heavier pipe. The added cost of this heavier pipe, when compared with alternative designs, might not result in the most economical overall system approach.

When pressure is increased at a booster station, it cannot be increased above the allowable operating pressure of the pipeline. Near the station, after the fluid leaves the pump or compressor, is the point at which operating pressure is highest in the segment of the pipeline between that station and the next. At that point, little of the energy supplied by the pump or compressor has been used to overcome friction.

Station equipment. Equipment in an individual station varies in both size and type, depending on the volume of fluid being handled and its properties, the size of the pipeline system, the type of monitoring and control used, the remoteness of the station, the environment, and other factors.

In a crude or products pumping station, the main items are pumps and their drivers. Many pumping stations have several pumps. The number depends on the total horsepower required and the individual capacity ratings of individual pumps. It can also depend on the designer's preference. It may be more desirable to have several smaller pumps, for example, than one large pump. This can provide flexibility of operation or a spare pump for use when one of the pumps fails or requires periodic maintenance.

Pump stations also typically include metering equipment for measuring throughput. Major stations, where custody of the fluid is transferred from one owner to another, contain a meter prover to calibrate the metering equipment. Originating stations may also have storage tanks to smooth out variations in flow to the station so the pumps will operate continuously at near-normal capacity, even though small changes in the supply of crude or products to the station occur. In addition to the main-line pumps at a station with storage, small booster pumps may be included to move liquid from storage tanks to the suction of the main-line pumps.

Many stations also include scraper traps. Scrapers, or *pigs*, are frequently run though pipelines for cleaning and other purposes. Points along the line, such as pump stations where piping manifolds are installed, make convenient points for inserting scrapers into the line and removing them. These scraper traps "catch" the pigs and allow them to be removed from the line without taking the entire pipeline out of service.

In addition to these key equipment items, a crude or products pump station often contains a complex array of piping and piping manifolds that permit the flow path to be directed to the pumps, to storage, or to other equipment. The manifolds also allow certain equipment to be bypassed when out of service for maintenance or repair. Installed in this piping system is a variety of valves and valve controllers. Some are operated manually; others operate automatically by monitoring flow conditions and reacting in a predetermined manner.

Natural gas pipeline systems also contain originating and booster stations. Rather than pumps, these stations contain gas compressors and their drivers. The selection of the number and size of these compressors is done based on criteria

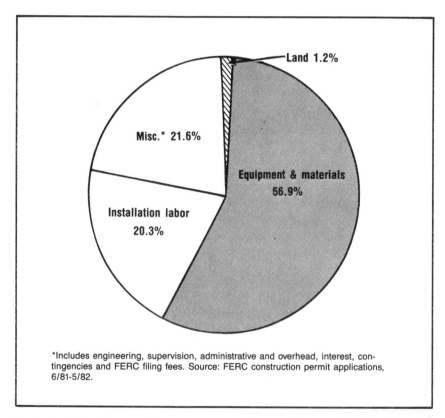

*Includes engineering, supervision, administrative and overhead, interest, contingencies and FERC filing fees. Source: FERC construction permit applications, 6/81-5/82.

Fig. 5–1.   United States onshore compressor station cost. Source: *Oil & Gas Journal*, 22 November 1982, p. 73.

similar to those used to choose the number and size of pumps in crude and products pump stations. And the same design philosophy may apply to determining whether to use a large compressor or several smaller compressors for the same thoughput volume. The cost of compressor stations is detailed in Figs. 5–1 and 5–2.

Compressor stations also contain measuring equipment, especially where ownership of the gas changes at the station. At some stations, typically originating stations, a separator is often required before the gas enters the compressor suction. This inlet separator removes liquids and sediments from the incoming gas stream to prevent damage to compression equipment. *Drips* are also used in natural gas compressor stations to remove liquids and sediment. The type and size of equipment for liquids removal depends on the amount of liquid contained in the incoming gas. Depending on the properties of the gas and compression conditions, other separators may also be required at the station to remove liquids condensed when the compressed gas is cooled.

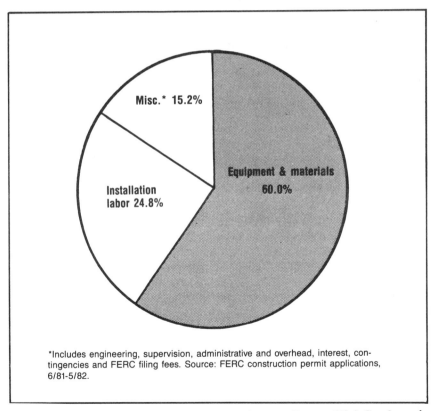

*Includes engineering, supervision, administrative and overhead, interest, contingencies and FERC filing fees. Source: FERC construction permit applications, 6/81-5/82.

Fig. 5–2.   United States offshore compressor station cost. Source: *Oil & Gas Journal,* 22 November 1982, p. 73.

Gas is usually compressed in stages, and the heat of compression must often be removed between compressor stages. Interstage coolers are used for this purpose.

Gas compressor stations also contain piping manifolds to direct the flow of gas entering and leaving the station, and valves and valve controllers to regulate flow.

The sophistication of the equipment in either a pump or compressor station depends in part on whether or not it is "attended" or "unattended." If an operator visits the station daily, for instance, the equipment in the station is generally less complex, and more valves and other equipment may be operated manually. If unattended for long periods and monitored from a central location, station equipment is generally more sophisticated and routine changes in valve position, for example, may be made automatically. The term *unattended* is relative; some stations are unattended for only the nighttime hours; others may not be visited for much longer periods.

Control equipment at pump and compressor stations also includes shutdown systems that automatically cause individual pumps or compressors—or the entire station—to be shut down if conditions exceed operating limits. For example, if rotating equipment overspeeds, automatic shutdown occurs; excessive pressure in the pipeline also may cause station shutdown. Other operating conditions, if exceeded, may also cause a shutdown.

Station safety regulations. In the United States, the Materials Transportation Bureau's pipeline safety regulations (49 CFR), discussed briefly in Chapter 1, contain requirements for compressor station design and construction, emergency shutdown systems, liquid removal, safety equipment, valves, and other aspects of compressor station design and operation. The location of buildings, building construction materials, building exits, and electrical facilities are specified.

Liquid removal facilities are required under these regulations at compressor stations when entrained vapors in the natural gas are expected to liquefy under anticipated pressure and temperature conditions to avoid those liquids entering the compressor and causing damage. Liquid separators must be equipped so liquids can be removed from the vessel manually; the vessel must be manufactured in accordance with the American Society of Mechanical Engineers' Boiler and Pressure Vessel Code. Liquid separators must also have either automatic liquid removal facilities, an automatic compressor shutdown device, or a high-liquid-level alarm.

In general, United States federal pipeline safety regulations require that, except for unattended field compressor stations of 1,000 hp or less, each compressor station must have an emergency shutdown system that does the following:

1. Is able to block gas out of the station and blow down the station piping
2. Will discharge gas from the blowdown piping at a location where the gas will not create a hazard
3. Provides means for the shutdown of gas compression equipment, gas fires, and electrical facilities in the vicinity of gas headers and in the compressor building
4. Can be operated from at least two locations, each of which is outside the gas area of the station, near exit gates or emergency exits, and not more than 500 ft from the limits of the station

Compressor stations that supply gas directly to a distribution system when no other source of gas is available must have an emergency shutdown system designed so that it will not function at the wrong time and cause unintended outages.

On offshore platforms, the emergency shutdown system for an unattended compressor station must be designed and installed to operate automatically when the gas pressure equals the maximum allowable operating pressure plus 15% and when an uncontrolled fire occurs on the platform. For a compressor station in a building, the emergency shutdown system must operate automatically when an uncontrolled fire occurs in the building or when the concentration of gas in the air reaches 50% or more of the lower explosive limit in a building that has a source of ignition.

United States pipeline safety regulations also require certain pressure-limiting devices to be installed in compressor stations. Each station must have pressure relief or other protection devices sensitive enough to insure that the maximum allowable operating pressure of the station piping and equipment is not exceeded by more than 10%. Each vent line that exhausts gas from the pressure relief valves must extend to a location where the gas can be discharged without hazard.

This is only an overview of United States federal pipeline safety regulations. More detail is contained in the regulations that is of critical importance to the pipeline system designer.

Pump application and design

In pumping any liquid, the goal is to add energy to the liquid to cause it to move through a pipeline by overcoming the resistance of friction and changes in elevation. Energy is supplied to the liquid through the pump by the pump's driver—either an engine, a turbine, or an electric motor. The amount of energy transferred from the pump's driver to the liquid depends primarily on the operating speed of the pump and the dimensions and geometry of the pump's liquid-handling parts. Pump efficiency, an overall indication of the energy transferred by the pump, depends on these and other factors.

Several types of pumps are used in handling crude and petroleum products. The capacity of each type can cover a wide range. The designer also has a variety of pump sizes and geometries to select from when choosing a pump for a specific application.

One basis for pump selection is the rating curve developed for each pump as a result of tests conducted by the manufacturer. Rating curves—also called efficiency curves and head-capacity curves—show how the pump's head, efficiency, and power consumption vary with its capacity.

The term *head* when used in pump design, refers to pressure or pressure differential. It comes from the use of a column of liquid (water is normally used as a standard, or base) to represent pressure. For instance, a column of water 10 ft high exerts a pressure of 4.335 psi at its base (1 ft of water exerts a pressure of 0.4335 psi). To pump water to the top of this column—to fill a tank, for instance—requires that the pump overcome a head of 10 ft, or 4.335 psi, disregarding friction losses. The head exerted by liquids other than water depends on their specific gravity. A crude oil with a specific gravity of 0.85 (specific gravity of water = 1) would exert a pressure of (0.85)(0.4335) = 0.368 psi/ft of height.

Each pump has an optimum operating range in which efficiency is a maximum. Ideally, a pump would be chosen that would operate within this range throughout its life. But in many cases, especially pipeline applications, capacity and other operating conditions may change significantly.

Two general types of pumps are common in oil industry applications: the centrifugal pump and the positive displacement pump. The choice of pump type depends primarily on the volume to be pumped and the pressure, or head, that must be overcome. In general, centrifugal pumps are used when the volume of liquid to be pumped is relatively large and pressures are moderate; positive displacement pumps are used for pumping smaller volumes at higher pressures. These application ranges are not sharply defined. The capabilities of each type overlap, and in many cases either could be used if pressure and volume capabilities were the only considerations.

The choice in such cases may be made by studying the rating curves of different pumps and determining which will operate most efficiently. Other factors, including compatibility with other system components and spare parts inventory, will also affect the pump type selected.

Positive displacement pumps. Positive displacement pumps are used for smaller volumes and when required pumping pressure is high. A common use for these pumps, also called piston pumps, is for injecting water into an oil-producing formation (waterflooding) to increase oil recovery. Pressures of several thousand pounds are often required to force the water down the injection well and into the formation.

Positive displacement pumps use a piston or plunger that is moved back and forth in a cylinder to increase the pressure of the liquid. The plunger is attached to a plunger rod, which in turn is connected to a crankshaft in the pump's power end. The crankshaft is driven by the pump's prime mover, an engine or electric motor.

Fluid enters the cylinder from the low-pressure source through suction valves. When the discharge valves are open, fluid is forced into the pipeline at high pressure. Single-acting plunger pumps fill the cylinder on the backward stroke and discharge fluid into the pipeline on the forward stroke. Double-acting reciprocating pumps fill the cylinder and discharge fluid on the same stroke. The cylinder on one side of the piston is filled as the other end of the cylinder is being discharged.

Single-acting positive displacement pumps are typically selected for use at high differential pressures and high operating speeds. Under these conditions, they may have a relatively short life. Double-acting plunger pumps are more suitable to applications involving high volumes, lower pressures, and slower speeds. Slow-speed operation can significantly prolong pump life.

Reciprocating pumps have been built with from one to several cylinders. A three-cylinder pump—a triplex—is common. Two-cylinder (duplex) pumps were used for many years on drilling rigs to pump drilling mud down the hole. Triplex pumps for that purpose are now common, and they are also used for other industry applications.

A consideration in choosing the number of cylinders a pump should have for a specific application is the pulsation caused by reciprocating pumps. The effect of the number of cylinders on pulsation can be seen by first considering a single-acting pump with one cylinder. The discharge pressure builds from zero (no flow) when the piston is on the filling stroke to the maximum discharge pressure when the piston is on the discharge stroke. The variation in discharge pressure and flow is great.

If the pump has two cylinders, however, one is discharging while the other is on the filling portion of the stroke. There still is variation in peak pressure, but the discharge pressure is never zero. Peaks in the pressure profile correspond to the discharge stroke of each piston. The difference between the peaks and valleys in this discharge pressure profile will be even smaller as the number of pistons increases further. The greater the number of pistons, the smoother the flow through the pump.

Uneven flow can be a problem in some applications. For instance, it causes pulsation in metering equipment that can affect measurement accuracy. Pulsation resulting from uneven flow can also cause vibration in piping and equipment that can result in failure. Fatique failure of piping due to pulsation-caused vibration can be serious, and the analysis and prevention of such vibration is complex. Excessive vibration is particularly undesirable on offshore platforms where vibration can be transmitted to structural elements of the platform. On land, when a pump is placed on a suitable foundation, the earth absorbs much of the vibration that could be present in the offshore structure.

Methods used to reduce or eliminate vibration caused by pulsation include bracing, use of flexible piping, and careful design of pump-connected piping. The pulsating flow will still exist because of the physical nature of the reciprocating pump; the goal is to control vibration so the pump and related equipment are not subject to damage or failure.

Vibration problems can be even more complex when more than one pump is discharging into a common piping system. The magnitude of the vibration may be increased above that which would occur with a single pump, depending on the frequency of vibration resulting from each pump and whether or not their pulsations are "in phase."

Efficiency.   Most reciprocating pumps have a packing element around the piston rod to seal the pump cylinder from the atmosphere. This packing can become ineffective due to normal wear or improper lubrication. When it is not sealing properly, air leaks into the pump cylinder, reducing pump efficiency.

Other causes of reduced efficiency in reciprocating pumps include the following:

1. Air or vapor in the suction line
2. Air or vapor above the suction valves
3. An air leak in the suction piping
4. Failure of valves to close properly
5. Worn valves and valve seats
6. Worn cylinders or plungers
7. Insufficient head (pressure) available at the pump suction

Plunger pumps are available with special corrosion-resistant metals (special "trim") for severe service applications.

Centrifugal pumps.   Centrifugal pumps (Fig. 5–3) are most common when large volumes must be pumped and high pressure differentials are not present. Rather than operating with a reciprocating motion, the centrifugal pump rotates. It consists of an impeller and a casing. The impeller is turned by the pump's driver through a shaft and "throws" the liquid into the pump casing, increasing the energy of the liquid by centrifugal force. This increase in energy causes the liquid to flow into the discharge line. Movement of the liquid out of the impeller reduces the pressure at the impeller inlet, allowing more fluid to flow into the impeller from the suction line.

Fig. 5–3. Centrifugal pumps for pipeline service. (Courtesy Byron Jackson Pump, Borg-Warner Corp.)

The result is a continuous flow of fluid through the pump. For continuous flow, however, these conditions must exist:[1]

1. Liquid must flow into the impeller at the same rate it is being discharged from the pump.

2. Pressure in the suction pump must exceed the pressure at the impeller inlet by an amount great enough to overcome suction line resistance and the difference in elevation, or lift, from the sump to the impeller.
3. Pressure inside the impeller cannot fall below the vapor pressure of the liquid.
4. Total head, or energy, developed by the impeller must be great enough to overcome the resistance of the system downstream of the pump.

There is a variety of types of centrifugal pumps, and capacities cover a wide range. Some pumps have one impeller and casing; others have several impellers and casings arranged in series. Impeller designs vary according to the manufacturer, the type of service, and operating conditions. Different casing designs also are available, and casings can be either one-piece or two-piece (split). Some centrifugal pumps are mounted along their centerline; inline pumps are mounted along a line between suction and discharge. Inline pumps are used primarily in plant applications rather than in pipeline service.

The location of the suction, or inlet to the pump, also distinguishes two types of centrifugal pumps. The inlet is located axially, concentric with the impeller, in end-suction pumps. In side-suction pumps, the suction inlet is perpendicular to the axis of the impeller.

In addition to capacity, centrifugal pumps are classified according to their specific speed, which relates flow rate, head, and operating speed of the pump.

Efficiency. As is the case with reciprocating pumps, key causes of reduced centrifugal pump efficiency are air leaks, and air or vapor pockets in the suction line or in the pump. Another important consideration in centrifugal pump operation is alignment of the pump with its driver. Improper alignment will cause vibration, overheating, and undue stresses on the shaft and other pump components. Proper alignment is a key step in pump installation. It must also be checked frequently during operation to ensure alignment is maintained.

In many pumping operations, cavitation is an important concern. Cavitation occurs when pressure decreases in a localized area of the pump. In some cases, the pressure may be reduced below the vapor pressure of the liquid, resulting in cavitation. Cavitation causes excessive vibration, reduced pump efficiency, and in some cases failure of pump components.

Net positive suction head. One of the most important criteria in designing a liquid pumping system is net positive suction head (NPSH). Each pump has its own requirement for NPSH, usually expressed in feet of head. The NPSH required is normally shown on the manufacturer's rating curve for each pump.

Net positive suction head available to a pump in a specific application must be equal to or greater than the required NPSH specified by the manufacturer. If enough suction head is not available at the desired flow rate, vapor lock, cavitation, and pump damage may result.

Basically, the net positive suction head is the difference in pressure between the tank or pipe from which liquid flows to the pump suction and the center of the pump suction. NPSH is determined from the pressure in the tank or pipe, the atmospheric pressure, the vapor pressure of the liquid, the specific gravity of the liquid, the friction losses in piping and valves, and the difference in elevation between the fluid in the tank and the pump suction.

A pump taking suction from a pressurized tank or vessel will have a greater NPSH available than if the vessel is at atmospheric pressure. And the higher the suction vessel is elevated above the pump, the greater will be the NPSH available.

NPSH for most pumps is calculated in the same way using these parameters. For reciprocating pumps, however, the motion of the pump piston must also be considered. An *acceleration head* is required to accelerate the fluid on each suction stroke so the fluid will "catch up" with the receding piston during the filling stroke.[2] If sufficient acceleration head is not provided, the pump will "knock" as a result of the fluid column catching up with the pump piston. Also, insufficient head to compensate for piston movement will cause cavitation due to incomplete filling of the cylinder. As in centrifugal pumps, cavitation reduces pump efficiency and can damage pump components.

Piping design.   The effect of NPSH on pump performance is so important that much effort goes into the proper design of suction piping, placement of the pump relative to the suction vessel, and selection of the valves required in the suction piping. Many of the design guidelines are aimed at reducing pressure loss due to friction in the piping.

Factors to be considered in the design of suction piping for a reciprocating pump, for example, include the following:[2]

1. The pump should be as close to the fluid supply as possible.
2. Use full-opening gate valves and avoid valves that constrict flow.
3. The ideal piping arrangement is short and direct, using no ells. If ells are required, use a 45° long radius instead of 90° ells.
4. If a reducer is required to change from one pipe diameter to a smaller pipe in the suction line between the main line and the pump, use an eccentric reducer rather than a concentric reducer. The straight side of the ecentric reducer should be on top.
5. Slope the suction line downward uniformly from the fluid supply to the pump to avoid air pockets.
6. If a bypass is installed, it should return liquid to the source vessel and not into the suction line.
7. Suction lines should be firmly anchored or buried to avoid strain on the pump and to help prevent vibrations from acting directly on the pump.

8. The suction line should never be smaller than the pump inlet. The line should be larger if possible. When two or more pumps are connected to a common suction header (section of pipe), proper sizing of that common header is important.
9. Install a properly sized pulsation dampener as close to the pump as possible.
10. Install a gate valve on the suction line to allow the pump to be isolated for maintenance.

Design of discharge piping for a pump installation is also important, and several guidelines are recommended.[2] A gate valve should be installed on the discharge line. By closing this valve and the gate valve on the suction line, the pump can be isolated. A check valve—a valve that permits flow in only one direction—is also recommended on the discharge line to prevent vibrations that might be caused by fluid flow from other units connected to common piping. A pulsation dampener on the discharge line is also recommended in many cases.

If possible, discharge lines should be run straight from the pump for at least 10 ft before a change in direction is made, and they should be securely anchored.

Many of these guidelines for reciprocating pump piping also are applicable to piping for centrifugal pumps. Pulsation and vibration, however, are less critical considerations in the design of centrifugal pump installations. Pulsation dampeners, for instance, are not usually needed for centrifugal pumps. Their rotary motion makes them less prone to significant pulsation or vibration caused by intermittent fluid flow. Vibration can, however, result from other causes.

The proper design of suction piping to provide adequate NPSH is just as important for centrifugal pumps as for reciprocating pumps.

Pressure and volume capability.   Key criteria in selecting a pump for a given application are the volume of liquid to be pumped and the differential pressure to be overcome. Differential pressure is the difference between the pressure at the suction of the pump and the required discharge pressure. Suction pressure is determined by the type and arrangement of the fluid source—pipeline, tank, or other vessel—and suction piping. Discharge pressure is that which is required to cause flow into the pipeline. Pressure in the pipeline depends on pipeline hydraulic conditions discussed earlier, the required pipeline delivery pressure, and the pressure loss between the delivery point and the pump discharge.

The pump selection process involves choosing a unit that will pump the required volume at the existing differential pressure when driven by a prime mover that will supply the required shaft horsepower.

As discussed earlier, manufacturers provide a head/capacity curve for each pump on which is plotted the volume the pump can discharge at a specified head, or *differential pressure*. Efficiency of the pump varies with head and flow volume. It is desirable to choose a pump that will operate at the highest efficiency under the design volume and pressure conditions.

Volume requirement can change substantially over the life of the system. It is important to consider expected changes in operating conditions and choose equipment that will operate efficiently over the life of the project. A number of pumps must be evaluated, and alternative arrangements of pumps or additions may have to be considered.

Several physical characteristics of pumps determine flow and head capacity. In reciprocating pumps, the larger the piston and cylinder, the greater the flow capacity. A longer stroke of the piston also increases flow capacity. Piston diameter and stroke length determine the volume of liquid taken into the cylinder on each stroke. In centrifugal pumps, the size of the impeller and case, and their geometry, affect the pump's capacity.

The combination of volume and pressure differential determines the energy that must be supplied to the liquid by the pump. For a given horsepower, the volume that can be pumped will decrease as the differential pressure increases.

The hydraulic horsepower that must be supplied by the pump is:

$$\text{HHP} = \frac{Q \times H \times \text{sp. gr.}}{3{,}960}$$

Where:

HHP = hydraulic horsepower
H = differential head, ft (pressure in psi multipled by 2.31 = head in ft)
Q = liquid flow, gal/min
sp. gr. = specific gravity of the liquid

To determine the horsepower that must be supplied by the pump driver, hydraulic horsepower must be corrected to account for pump efficiency. For instance, if pump efficiency is 85%, the horsepower that must be supplied by the pump's driver is HHP/0.85.

In addition to supplying the required differential pressure, a pump must be chosen that can safely withstand the absolute discharge pressure. For example, though two pumps might both be capable of overcoming a pressure differential of 100 psi, a pump to handle the differential pressure between 100 and 200 psi would not be suitable where suction and discharge pressures are 2,000 and 2,100 psi, respectively.

In many cases, more than one pump is required at a station. Several pumps can be connected in different ways to provide a range of operating conditions and capabilities. In a parallel arrangement (Fig. 5–4), more than one pump takes suction from a single source. A suction manifold, consisting of a pipe from which individual suction lines branch off to the inlet of each pump, is used in this case. Then each pump discharges separately into a discharge manifold connected to the pipeline. When connected in parallel, each pump operates at approximately the same suction and discharge pressure, and the total flow volume is the sum of the output of the individual pumps.

Pumps can also be connected in series (Fig. 5 – 4). In this case, one pump takes suction from the fluid source, then discharges to the suction of a second pump. The last pump in series discharges into the pipeline. The suction pressure for the second pump is equal to the discharge pressure of the first pump minus losses in the connecting piping. In this arrangement, the total flow volume is handled by each pump, but the total differential head is the sum of the differential heads of the individual pumps.

Parallel operation is represented by a situation in which pipeline flow increases enough to require an additional pump at a station. The new pump is installed alongside an existing pump and the suction and discharge manifold are extended to connect the new unit.

An example of series pump operation is an installation in which large main line pumps are fed by smaller pumps that take suction from a storage tank. The small suction booster pumps handle a large volume at a relatively low differential head. They supply the required net positive suction head to the main line pumps, which would not be available if the large pumps were connected directly to the storage tank.

Though some data are not required for some pump applications and types, most of the following information is needed to properly select and size a pipeline pump:

1. Characteristics of the fluid, including specific gravity at pumping temperature, pumping temperature, vapor pressure at pumping temperature, and the presence of any corrosive materials
2. Desired pumping rate and expected future changes in volume requirements
3. Pressure conditions, including suction and discharge pressure, net positive suction head available, expected future pressure conditions, and whether the pump will operate in series or parallel with other pumps
4. Preferred type of pump and type of shaft seal for centrifugal pumps
5. Special metallurgy required to handle high temperature, corrosive fluid, or other severe conditions
6. Type of pump driver to be used and any space limitations

An economic evaluation of possible alternatives is an important part of pump selection. Operating costs and initial capital costs must be compared for each alternative. Other factors to consider include annual maintenance costs and fuel efficiency; expected increases in operating, maintenance, and fuel costs; and special limitations or advantages of each pump. The choice cannot often be

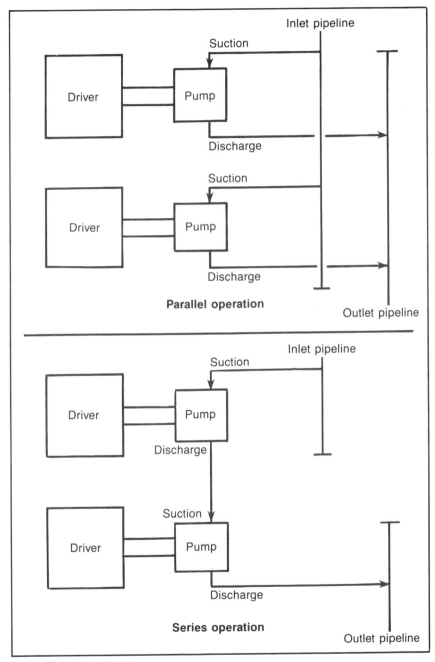

Fig. 5-4. Parallel pump arrangement (top) and series arrangement (bottom).

made on any one of these criteria; it must be based on the best combination of features for a specific application.

Selecting the proper pump is only a part of pump station design. The other major equipment item, the pump driver, will be discussed in more detail in Chapter 6. Pump station design also involves liquid metering devices, meter provers, scraper traps, and storage vessels. Station piping is an important design element, especially where piping pressure losses and net positive suction head are critical. Selecting valves to control fluid flow is a key to the operation of individual pumps and the station as a whole.

Station monitoring and control is critical to reliable operation of the equipment. Individual pumps may be designed to shut down automatically when high temperature, low flow, excessive pressure, high lube oil temperature, or other conditions occur. Shut-down systems are designed to stop a pump before severe damage occurs. Though damage is avoided, the cost of down time can be great. In many cases, this cost can justify the installation of standby units. A standby pump can be put on line quickly if a pump is shut down and can be used when a pump must be taken out of service for maintenance.

Compressor application and design

Pumps and pump stations for liquid pipelines have much in common with compressors and compressor stations for natural gas pipelines. The key difference is the fluid being handled: liquids are incompressible and gases are compressible.

Compressors, like pumps, add energy to the flowing fluid to cause it to move through the pipeline. Like pumps, compressors can be divided generally into reciprocating (Fig. 5–5) and centrifugal units (Fig. 5–6). Generalizations made for reciprocating and centrifugal pumps also can be made for these two types of compressors. For instance, reciprocating compressors generally operate at slower speeds than centrifugal units and are used in applications where relatively high pressures are required.

As is the case with positive displacement pumps, reciprocating compressors also produce a pulsating flow. A reciprocating compressor installation must be designed to avoid equipment and piping damage resulting from pulsation and vibration.

Reciprocating compressors. Many reciprocating compressor units used for natural gas pipeline service are integral, i.e., the compressor driver and the compressor are contained in a single unit. In a large multicylinder compressor, several compressor cylinders and the engine cylinders are both connected to the same crankshaft. Natural gas is used for engine fuel. As the engine rotates the crankshaft, rods connecting the crankshaft to compressor pistons reciprocate the pistons in the compressor's cylinders. In a typical machine, the engine cylinders are either

Fig. 5–5. Integral engine/compressor, rated at 11,000 bhp, has 20 power cylinders in V configuration (background) and ten compressor cylinders (foreground). (Courtesy Dresser Industries Inc., Dresser Clark Division.)

vertical or in a vertical V configuration and the compressor cylinders are horizontal.

There are also reciprocating compressors that do not have integral drivers. These compressors are generally smaller than integral machines and are often used for auxiliary services.

If the compressor cylinders of a single integral reciprocating compressor unit are operated in parallel, each cylinder compresses a portion of the total gas volume and each cylinder operates with approximately the same suction and discharge pressures. It is not uncommon, however, for one compressor unit to include more than one stage of compression by connecting individual compressor cylinders in series. In this configuration, each cylinder handles the total volume, but the discharge pressure of the first-stage cylinder is equal to the suction pressure—less piping losses—of the next cylinder. If a large total compression ratio exists, this series arrangement can be used.

Reciprocating compressor cylinders contain suction and discharge valves to permit the flow of gas into and out of the cylinder. Gas flowing into the cylinder through the suction valve at suction temperature and pressure is compressed in the cylinder and discharged at a higher pressure through the discharge valve.

The volume that the unit can compress under given pressure conditions depends on the size of the cylinder, the length of the piston stroke (cylinder size and stroke length determine piston displacement), and the clearance volume

Fig. 5–6.   Axial centrifugal compressor, rated at 27,400 bhp, is driven by gas turbine (in white enclosure behind compressor). Inlet is from right. (Courtesy Dresser Industries Inc., Dresser Clark Division.)

within the cylinder. Clearance volume is the volume remaining in the compressor cylinder at the end of the piston's discharge stroke. It is the volume between the end of the piston and the end of the cylinder plus the volumes contained in valve ports and other areas. Clearance volume is usually expressed as a percent of piston displacement.

Compressor cylinders operating at low pressure generally are larger than those handling an equal volume of gas at a higher pressure. As gas is compressed, the same amount of gas occupies a smaller volume. On an integral reciprocating compressor in which the cylinders are arranged in series, the size of the compressor cylinder decreases as the pressure increases.

Compressor manufacturers offer a number of standard compressor frames. The proper compressor cylinders are mounted on one of these standard frames to meet a specific application. Each frame has a maximum speed and load-carrying capacity; load-carrying capacity involves horsepower and compressor-rod loading. These ratings are established according to the permissible loads on the crankshaft, connecting rod, and other components, and the ability of bearings to dissipate heat.

It is generally considered desirable to have a reversal of rod loading during each stroke. This allows bearing surfaces to part on each stroke so they can be properly lubricated.[2] Reciprocating compressors and their drivers also require cooling and lubrication systems to prevent excessive buildup of heat and damage to pistons and cylinders.

Centrifugal compressors. In centrifugal compressors, energy is added to the gas by the rotation of an impeller rather than by confining the gas and then compressing it, as is the case in a reciprocating compressor. A centrifugal compressor discharges gas at high velocity into a diffuser, where gas velocity is reduced and its kinetic energy converted to pressure.

Centrifugal compressors consist of a housing, an impeller mounted on a rotating shaft, bearings, and seals to prevent gas from escaping along the shaft. The shape and size of the diffuser and impeller vary, depending on operating conditions and on the manufacturer.

Centrifugal compressors have fewer moving parts than reciprocating units. Only the shaft and impeller of a centrifugal unit rotate, while reciprocating compressors contain connecting rods, bearings, and other components to convert crankshaft rotation to reciprocating motion. Hence, centrifugal compressors generally have lower maintenance costs and lower lubrication oil consumption.

The output from a centrifugal compressor is smooth compared with the pulsating flow of reciprocating compressors. Because of this feature, they are often considered for installation on offshore platforms where vibration must be minimized. Special efforts have been made in recent years to extend the operating pressure range of centrifugal compressors to meet the demands of offshore applications. However, since centrifugal compressors operate at relatively high speeds, they must be properly balanced.

Centrifugal compressors are not capable of as high a compression ratio as reciprocating machines, but they can be arranged in series so each is only required to develop a portion of the total differential pressure required. Their continuous flow characteristics make this series arrangement practical.

Multistage centrifugal compressors, basically a series of single-stage compressors contained in a single casing, are also used.

Compression ratio. A key consideration in designing any compressor installation is the compression ratio. If the overall compression ratio is high, several compressor stages may be required. Compression ratio must be limited to avoid excessive temperatures that would result from too high a ratio. When gas is compressed, its temperature increases. The more it is compressed, the greater the temperature increase. Recommended operating temperatures for compressor cylinders are used as a guide in determining maximum compression ratio per stage.

Compression ratio required per stage is calculated using the overall ratio of the installation and pressure losses in suction and interstage cooling piping. For example, a close estimate of the compression ratio per stage for a two-stage compressor can be obtained by calculating the square root of the station discharge pressure divided by the station suction pressure. If it is necessary to compress gas from 100 psia to 900 psia with a two-stage compressor, the compression ratio per stage is approximately:

$$\sqrt{900/100} \ = \ \sqrt{9} \ = \ 3$$

If a three-stage compressor is to be used, the cube root of the ratio of station discharge pressure to station suction pressure gives the approximate compression ratio per stage:

$$\sqrt[3]{900/100} \ = \ \sqrt[3]{9} \ = \ 2.08$$

If three stages are used (disregarding suction piping and interstage cooling pressure losses), the first stage would take suction at 100 psi and discharge at 208 psi (100 × 2.08). The second stage would take suction at 208 psi and discharge at 208 × 2.08 = 433 psi. The final stage suction would be at 433 psi and would discharge at the desired pressure of 900 psi (433 × 2.08).

For final design, suction losses and losses between stages must be considered. This will change the estimated compression ratios slightly, but they can be easily recalculated using the actual suction and discharge pressures and compared with any limits. Maximum allowable compression per stage is usually based on recommended limits on compression ratio or operating temperature.

The temperature increase that will result from compression from a given suction pressure to a given discharge pressure can be calculated using the temperature of the gas at suction conditions, the suction and discharge pressures, and the heat capacity of the gas. The calculated discharge temperature can then be compared with recommended limits.

To limit gas temperature to recommended values, it is often necessary to cool the gas between compression stages. Interstage cooling between compression stages can be done by air cooling, by cooling with water in a heat exchanger, or by exchanging heat with the inlet gas in a gas-to-gas heat exchanger.

Typically, air coolers consist of banks of horizontal coils through which the gas flows between compression stages. A fan blows air over the coils to cool the gas before it enters the suction of the next compressor stage. In exchanging heat with inlet gas or with water, other types of equipment are used in which one fluid flows within the coils, or tubes, and the other fluid flows on the outside of the tubes.

Depending on the type of gas, separators may also be needed between compression stages to remove any liquids condensed by interstage cooling. If liquids enter the compressor, damage can result. In reciprocating compressors, damage can occur because the liquid is not compressible and light hydrocarbon liquids may "wash" some of the lubricant from the cylinder wall. In centrifugal compressors, liquid droplets impinging on the rapidly rotating impeller at high velocity can cause damage.

Capacity, horsepower.   Compressor capacity can be expressed in several ways. A common method is to express capacity in cubic feet per unit of time at suction temperature and pressure. Capacity is also calculated at base, or standard, conditions of temperature and pressure. Because gases are highly compressible, their volume changes directly with temperature and pressure. At constant pressure, volume increases with increasing temperature; if temperature is held constant, volume decreases with increasing pressure.

As discussed in Chapter 4, a compressibility factor must be included in volume calculations to account for the deviation of a gas from the ideal gas law. Because natural gas is a mixture of several components with different physical properties, the properties of the mixture must be determined from a gas analysis, as was described in Chapter 4, for accurate calculation.

The volume that a given reciprocating compressor can handle depends on the piston displacement and the volumetric efficiency of the cylinder. Piston displacement depends on whether or not the piston is single acting (compresses only on one end of the cylinder) or double acting. *Piston displacement* is the net piston area multiplied by the length of the piston stroke. Including the speed of the compressor in the calculation then gives total displacement over a unit of time. In a single-acting unit, the piston area is the area of the end of the piston. In a double-acting machine, the area occupied by the piston rod must be subtracted from the area of the piston on the rod end to give the net piston area. Volumetric efficiency, discussed earlier, is also required in reciprocating compressor capacity calculations.

The capacity of a centrifugal compressor depends on the size and speed of its impeller and the pressure against which the compressor is discharging. Centrifugal compressor capacity varies directly with speed. Manufacturer's charts are available that tell how volume, head, and compressor speed are related.

The typical design problem is not to determine the capacity of a compressor but to select a compressor to handle a given volume of a specific gas at the prescribed pressure conditions. To do this, it is necessary to determine how much compressor horsepower will be required to handle the required volume. Several approaches to calculating required horsepower are possible, depending on the type of compressor and operating conditions. Factors that affect the brake horsepower required include the volume to be compressed, suction and discharge pressures (or compression ratio), the heat capacity of the gas, and the efficiency of the compressor.

For both reciprocating and centrifugal compressors, charts have been developed that show horsepower required to compress 1 MMcfd of a gas with a given ratio of specific heats at various compression ratios. These charts are used for a preliminary selection of compressor sizes. Then calculations are made to confirm that the compressor is capable of handling the required volume.

For instance, if it is necessary to compress 10 MMcfd of a gas with a specific heat ratio of 1.2 from 14.7 psi to 30 psi with a single-stage reciprocating compressor, one chart indicates about 43 brake horsepower per MMcfd is required. For suction pressures other than 14.7 psi, it is necessary to correct the brake horsepower per MMcfd for the actual suction pressure. In this example, total compressor horsepower required is determined by multiplying the brake horsepower per MMcfd by the total volume to be compressed:

$$\text{total BHP} = (43 \text{ BHP/MMcfd}) \times (10 \text{ MMcfd}) = 430 \text{ BHP}$$

Much more rigorous methods of calculating compressor capacity and horsepower can be used. For a complex installation where (1) volumes will change over time, (2) several fuel options must be considered, and (3) other variables exist, the selection of natural gas compressors is much more sophisticated. In the final evaluation, accurate values for the physical properties of the gas mixture, supercompressibility factors, and other variables are needed.

The job, however, is one of compressor selection rather than design. The pipeline engineer does not often design a unique compressor for his specific application but chooses from among those available from manufacturers.

As in the case of pumps, the best compressor for each installation is the one that represents the most desirable combination of capital cost; annual operating and maintenance cost; fuel efficiency; expected increases in operating, maintenance, and fuel costs; and the specific advantages and limitations of each alternative.

Other considerations.   This general outline of the considerations involved in selecting compressors is only a part of compressor station design. The other major equipment item in a compressor station, the compressor driver, will be discussed in Chapter 6.

Considerable time and expertise must also be devoted to providing auxiliary services, including lubrication, cooling, monitoring and control instrumentation, interstage cooling, and liquid removal. Compressor lubrication equipment must be reliable. It often must operate in severe environments where dust, extreme cold, or other conditions exist. Though compressors are protected by sophisticated shut-down systems, resulting downtime is costly. And if the compressor is not shut down when the lube system fails, severe damage results. Cooling systems must also be designed for reliability and safety.

The design of interstage cooling involves sizing heat exchangers by calculating heat flow and the sizing of vessels if liquid removal is required. Additional design work may involve pulsation dampeners, the selection of valves to regulate and distribute flow, and station piping.

REFERENCES

1. S. Yedidiah, *Centrifugal Pump Problems: Causes and Cures,* Tulsa: Penn-Well, 1980.
2. *Engineering Data Book,* ninth edition, Gas Processors Suppliers Association, 1972.

6

PRIME MOVERS

THE purpose of a pipeline prime mover—an engine, motor or turbine—is to supply the shaft horsepower required by a pump or compressor to move fluid through the pipeline.

Prime movers are not designed and manufactured for each specific application. Rather, manufacturers make available a number of sizes of different types of drivers that can be applied over a relatively wide range of operating conditions. Options or modifications may be available to increase a unit's operating range. The pipeline designer's task is to choose the most suitable type of driver to be used and the proper machine within that type.

Basic selection parameters are horsepower output and efficiency. Several considerations, in addition to horsepower rating, are important in selecting pump or compressor drivers. Because the designer wants the most economical installation, he must also consider the availability of energy to power the prime mover and the cost of energy. Initial cost of the unit must be compared with that of other units that could provide the required output horsepower, and maintenance costs must be estimated.

One of the most difficult steps in an economic evaluation of various prime mover choices is the projection of costs over the life of the project or unit. The cost of fuel is an example of the difficulty in predicting operating costs. Fuel costs can rise rapidly, as they did beginning in 1973. But the cost of different fuels can also change relative to each other. The cost of one source of energy may even decline during part of the project's life.

As difficult as it is, a good estimate of fuel and other operating costs is necessary to compare prime mover alternatives accurately. For example, a machine with the lowest initial cost may have an operating and maintenance cost over the project's life that will make it less economical than another type of prime mover with a higher initial cost. As in most economic evaluations, the time value of money is an important basis by which to compare alternatives.

Choosing the type of driver

Types of prime movers include electric motors, gas turbines, and diesel and internal combustion engines. Diesel engines are not widely used for main line service; electric motors and gas turbines have become the most popular for this application.

The type of energy available to power the installation is the key consideration in choosing among driver types. If electric power is not available, for instance, without building a long electric line to the site and natural gas is available nearby, an electric motor driver will likely not be the most economical choice. If a relatively long gas line must be built, then the cost of using a gas turbine or natural gas engine may be excessive. Reliability of the energy source is another important factor. If electrical outages are common in the system to be used or a long-term gas commitment cannot be made, the choice of pump or compressor prime mover will be influenced. Frequent shutdowns resulting from interruptions in the supply of energy or fuel are costly.

Availability and dependability of the energy source, in addition to the projected cost of fuel or electricity over the life of the project, must be evaluated.

In comparing cost, fuel use, maintenance expense, and other features, it is common to use a unit basis. For instance, the initial cost of a prime mover is typically expressed in dollars per brake horsepower ($/bhp) when comparing units with different horsepower output ratings. Fuel consumption is often expressed in lb/bhp/hr for diesel engines or BTU/bhp/hr for natural gas engines. Maintenance costs are often expressed in $/bhp/yr when comparing different units. In comparing initial cost, the basic prime mover and all necessary auxiliary equipment must be considered.

Other factors influence the choice of prime mover type, including other units in a company's system. If a large portion of a pipeline company's pumps or compressors are driven by one type of prime mover, there are advantages in continuing to use that type, providing cost and suitability are otherwise acceptable. Maintenance and operating personnel may require additional training if a different type of prime mover is chosen, and an additional spare parts inventory will be required.

The type of supervisory control system to be used at the pump or compressor station and the pipeline's overall monitoring and control configuration are also considered when selecting prime movers. In the past, some control functions have been more easily performed with one type of prime mover than with another, but today's sophisticated instrumentation makes it possible to monitor and control almost any parameter on any type of driver.

Recommended operating speed of the pump or compressor to be driven also influences the type of driver. The prime mover speed must be compatible with the speed of the machine to be driven. If a gas turbine is used, it is often necessary to use a speed reducer between the gas turbine and the pump or compressor to match the driver's shaft speed to that of the driven unit.

Other considerations in choosing the proper type of prime mover include efficiency, availability, and expected time between major inspections and overhauls. Table 6–1 is an example comparison of prime mover alternatives.

There is no average compressor or pump station size, nor average pump or compressor size within a station. A small gas-gathering system compressor station can have one compressor unit within the station; that compressor's rating may be as small as 100 hp. From there, size ranges upward to large main line transmission or pumping stations with a number of individual units totaling as much as 30,000 hp or more.

A large system with many compressors can include prime movers with horsepower outputs totaling several hundred thousand horsepower. For instance, Texas Gas Transmission Corp.'s 5,900-mile gas transmission system includes compressor engines that develop a total of about 500,000 hp. The system includes 29 compressor stations. One of the largest companies has over 27,000 miles of field-gathering and transmission pipeline that is powered by 205 compressor stations.[1]

Horsepower required. To be considered for a specific application, a prime mover must be capable of supplying the shaft horsepower required by the pump or compressor. Then those drivers that meet this basic criterion are further evaluated on the basis of initial cost, fuel cost, maintenance and operating cost, control adaptability, flexibility of operation, and other factors.

Horsepower required to drive liquids pumps is determined by first calculating the hydraulic horsepower the pump must deliver. Hydraulic horsepower is then divided by pump efficiency to determine the brake horsepower that must be input to the pump shaft. The power that must be input to the pump shaft is the shaft horsepower that the prime mover must deliver. Obviously, the higher the pump efficiency—the lower are friction and other losses within the pump—the less prime mover horsepower is required to move a given volume of liquid through the pipeline.

The hydraulic horsepower that a pump must supply to the fluid is the energy required to overcome elevation and friction losses. It is calculated using this equation:

TABLE 6-1
Comparison Of Prime Movers*

	Electric motor	High-speed diesel (900–1,800 rpm)	Low-speed diesel (400–600 rpm)	Aircraft turbine	Light Air/ind. turbine	Industrial turbine
Initial cost, $/bhp	20–30	100–150	175–250	150–200	150–200	170–250
Fuel rate	(Energy rate + demand)					
A. No. 2 diesel lb/bhp/hr		0.48	0.59	0.51	0.51	0.59
B. Gas, BTU/bhp/hr		8–10,000	10–12,000	9–10,000	9–10,000	10–12,000
Maintenance,† $/bhp/yr	0.75	25–45	5–10	20–30	20–30	5–10
Efficiency,‡ %	95	35	30	26	24	22
Speed, rpm	1,200/1,800/3,600	900–1,800	400–600	8,000–30,000	14,000–25,000	6,000–8,000
Availability on load base, %	99.9	90	99	95	97	99
Time between overhauls, hr	100,000	30,000	75,000	25,000	30,000	100,000
Time between major inspections, hr	25,000	4,000	20,000	6,000	10,000	30,000

*1,000 to 5,000 bhp in base-load operation. †Maintenance includes all parts and contract labor (excludes fuel) for U.S.A. onshore. ‡Turbine overall efficiency can be doubled or tripled by applying heat recovery.

Source: Oil & Gas Journal (**16 February 1981**), p. 87

$$HHP = \frac{H \times Q}{33,000}$$

Where:

HHP = hydraulic horsepower
H = head in ft that must be overcome (elevation and friction)
Q = liquid flow, lb/min
33,000 = conversion from ft-lb/min to hp (1 hp = 33,000 ft-lb/min)

This equation is valid when pumping water. When pumping other liquids, the specific gravity of the liquid must be included in the equation. The lower the specific gravity of the fluid being pumped, the less horsepower is required for pumping a given volume.

Flow in the pipeline industry is normally given in volume units rather than weight units as in the equation above. The following equation uses volume units and includes the effect of fluid specific gravity and pump efficiency. The equation determines the brake horsepower that must be supplied to the pump shaft by the prime mover.

$$bhp = \frac{H \times Q \times sp.\ gr.}{3,960 \times eff.}$$

Where:

bhp = brake horsepower required
Q = liquid flow, gal/min
H = head, ft, that must be overcome (elevation and friction)
sp. gr. = specific gravity of the fluid (in liquids pipelines, this may range from 0.35 for ethane to 0.95 for fuel oils)
3,960 = conversion from ft-gal/min to horsepower
(1 hp = 33,000 ft-lb/min and 1 gal = 8.33 lb; therefore, $\frac{33,000\ ft\text{-}lb/min}{8.33\ lb/gal}$ = 3,960 ft-gal/min = 1 hp)

To see the effect of pump efficiency on prime mover size, assume two pumps are being evaluated for pumping a crude oil with a specific gravity of 0.90. Of the two pumps being considered, one has an efficiency of 80% (0.80) and one an efficiency of 70% (0.70). The brake horsepower required to pump 100 gal of crude against a head of 100 ft with the more efficient pump is:

$$bhp = \frac{100 \times 100 \times 0.90}{3,960 \times 0.80} = 2.84\ hp$$

For the same fluid and pumping conditions, the pump with the lower efficiency requires:

$$bhp = \frac{100 \times 100 \times 0.90}{3,960 \times 0.70} = 3.25\ hp$$

A typical crude or natural gas liquids pipeline transports volumes of several thousand b/d, and friction and elevation head may total several hundred feet. Under these conditions, the 14% difference in horsepower requirements in this example is significant.

Sizing compressor drivers.   Calculating the amount of horsepower required to drive a compressor involves determining the theoretical horsepower required to increase gas pressure from suction pressure to discharge pressure, then allowing for losses in the compressor. Because gas is a compressible fluid, more terms must be used in the calculation than in calculating hydraulic horsepower for pumping.

Assuming adiabatic compression (compression in which no cooling occurs and gas temperature rises steadily), theoretical horsepower for a reciprocating compressor, for example, can be calculated from this formula:

$$\text{Theoretical hp} = \frac{P_1 V_1}{229} \times \frac{k}{k-1} \left[\left(\frac{P_2}{P_1} \right)^{(k-1)/k} - 1 \right]$$

Where:

k = ratio of specific heats of the gas (c_p/c_v)
P_1 = suction pressure, psia
P_2 = discharge pressure, psia
V_1 = suction volume, cu ft/min

Actual horsepower required is then obtained by multiplying theoretical horsepower by a factor that accounts for losses due to pressure drop through valves and piping, and the friction of piston rings and rod packing.[2]

The brake hp/MMcfd is the brake horsepower required to compress 1 million cubic feet of gas per day; the gas is measured at 14.4 psia and suction temperature. Charts are available for estimating bhp/MMcfd quickly if the compression ratio and the specific heats of the gas are known. These charts can be used to estimate total horsepower required by multiplying bhp/MMcfd from the chart by the volume of gas to be compressed, in MMcfd. In the final evaluation of a compressor drive, more detailed calculations may be required.

Though the equation may take a slightly different form, the same approach is used in calculating the horsepower required to operate a centrifugal compressor. Theoretical, or gas, horsepower is determined. Then a factor to account for losses in the compressor is used to calculate the horsepower the driver must supply to the compressor shaft. Horsepower calculations for centrifugal compressors also take into account volume to be compressed, suction and discharge temperatures and pressures, and the ratio of specific heats. For most centrifugal compressors, mechanical losses are relatively small.

Electric motors

The key consideration in selecting electric motors for a pipeline prime mover application is often the cost and availability of electric power. The initial cost of an electric motor is often lower than that of other prime movers, provided it is not necessary to build a long electric transmission line.

In general, electric motors are easily adapted to automatic control and remote operation, and maintenance requirements are relatively low. Electric motors are not normally used to drive reciprocating compressors except small units used for auxiliary services, such as supplying instrument air. The operating speed of electric motors is not usually compatible with the operating speed of reciprocating compressors.

Electric motors are used to drive both centrifugal compressors and centrifugal pumps (Fig. 6–1). Electric prime movers for main line pipeline service are generally induction squirrel-cage motors, although synchronous motors are used for special service.[3] These main line motors are generally three-phase, high voltage motors with enclosures to provide protection from weather.

Fig. 6–1. Electric-motor-driven centrifugal pump. (Courtesy Ingersoll-Rand Company.)

An induction, or squirrel-cage, motor has a stator consisting of three sets of phase coils within a laminated steel core supported within a frame. When the stator is energized with three-phase alternating current (AC) voltage it establishes a rotating electric field that acts like a magnet. The rotating motor component, when exposed to the rotating electric field, becomes a magnet that tries to follow the rotating electric field of the stator.

The amount of force between the rotating field and the rotor determines the motor's torque. The motor must supply enough torque to turn the shaft of the pump or compressor. Torque required when the motor is started differs from the torque requirement after the pump is running. Both starting and full-load torque must be considered in sizing and selecting electric prime movers.

Synchronous electric motors are also used for main line pumping service. A synchronous motor has a rotor that is energized with direct current (DC) voltage. The amount of voltage supplied to the motor determines how strong the magnetic field,of the rotor will be and the amount of output torque.

Electric prime movers for pumping service normally operate at speeds of 1,800 rpm or 3,600 rpm, though motors that operate at other speeds are available.

The manufacturer supplies curves with each motor model that indicate the motor's characteristics, including speed vs current, speed vs torque, speed vs efficiency, and speed vs power factor.[4] These curves are representative, but they do not guarantee the motor will perform exactly as the curves indicate. Tests can be run and curves provided, however, that do guarantee motor performance. There is normally an added cost for such testing.

In selecting electric prime movers, it is important to choose a motor with the highest efficiency and the highest power factor possible. The power factor accounts for the fact that coils in an electric motor cause electricity to act differently than, say, in a light bulb. The result is that above the theoretical power calculated by multiplying volts by amperes, additional power must be supplied to the motor. The power factor of a light bulb could be considered to be 1, but the power factor of a motor is less than 1. Because additional power must be supplied to the motor, the cost of energy to power the motor increases as the power factor decreases. The power factor of a motor can be made to approach 1 by proper design.

Electric motors can pose a hazard in explosive environments, such as those that can exist around a pump or compressor handling oil, other petroleum liquids, or natural gas. Therefore, a variety of motor enclosures is available. Various industry codes specify certain zones or areas in which a specific type of electric motor enclosure must be used. The open-type enclosure is the least expensive; an explosion-proof enclosure is the most costly. Types of enclosures include the following:[5]

1. In open-type enclosures, the stator is supported by an open frame that includes bearing pedestals to support the rotor.
2. Dripproof and splashproof motors are similar to the open type but have a shield to prevent water from entering the motor from overhead (dripproof) or from the sides as a result of splashing from the base (splashproof).

3. Totally enclosed fan-cooled (TEFC) motors have internal fans to circulate air to dissipate heat generated by the motor.
4. Explosion-proof motors are also completely enclosed with an internal fan to circulate air. This enclosure is capable of withstanding an internal explosion without permitting hot gases to escape that could create an explosion outside the case. This is the heaviest and most expensive electric motor enclosure.

Other considerations in selecting an electric prime mover include insulation, physical specifications of the motor and mounting arrangement, the need for air filter equipment, and bearing type. The operating environment will be a factor in choosing the type enclosure, type insulation, and other options. Abrasive, high-moisture, chemically corrosive, and other severe environments may require special options.

Economics.   As mentioned earlier, the initial cost of an electric prime mover is typically less than that of a gas turbine or reciprocating engine driver. Wright estimates the initial basic cost of an electric motor for pumping services as $20-$30/bhp, compared with $150-$250/bhp for various types of gas turbines.[6]

Maintenance cost for electric motors was estimated in the same report at about $0.75/bhp/year, considerably less than the estimated maintenance cost of gas turbines. The cost of maintaining a given driver is determined by multiplying this unit cost by the prime mover's rated horsepower. For instance, using this estimate, maintenance cost for a 1,000-hp electric motor would be $750/year.

Today, energy is an important cost item in comparing prime mover choices. In the case of electric motors, the cost of power depends on the power demand (kilowatts), the length of time that power is used (kilowatt-hours), and the power factor. For example, assume a 1,000-hp electric motor operated 500 hours during one month. If motor efficiency is assumed to be 0.95 the power demand would be:

$$\frac{(1,000 \text{ hp}) \times (746 \text{ w/hp})}{0.95} = 785,260 \text{ w} = 785 \text{ kw}$$

The amount of energy used during the month would be:

$$(785 \text{ kw}) \times (500 \text{ hr}) = 392,500 \text{ kw-hours}$$

Typically, the pipeline company would be billed by the power company at a specified charge for demand plus a kilowatt-hour rate for the amount of energy used. The cost of energy is also adjusted according to the prime mover's power factor. The unit cost/kw-hour may be on a sliding scale, decreasing as the amount of energy used increases above a certain level. For example, the first 40,000 kw-hours might cost 1.75¢/kw-hour; the next 160,000 kw-hours might cost only 1.20¢/kw-hour.

Additional adjustments may also be made to the cost of electric service including on-peak or off-peak charges to encourage use of power during other than peak periods and minimum demand charges.

Gas turbines

Gas turbine prime movers are widely used to drive pipeline pumps (Fig. 6–2) and centrifugal compressors (Fig. 6–3) in natural gas service. They are not normally used with reciprocating natural gas compressors because of the difficulty in matching turbine speed to compressor speed.

A variety of gas turbines is available for pipeline service. Generally, the industrial turbine is a moderate speed machine, typically operating at

Fig. 6–2.   Gas turbine inside housing drives crude pipeline pumps. Source: *Oil & Gas Journal*, 14 March 1977, p. 67.

Fig. 6–3.   Turbine engine drives natural gas compressor. Source: *Oil & Gas Journal,* 23 July 1979, p. 43.

6,000–8,000 rpm. Another type of turbine whose application has grown in pipeline service is the aircraft-derivative turbine. These units were originally designed for aircraft power but have been modified for gas compression service with considerable success. Aircraft-derivative turbines operate at high speeds, typically in the 8,000–30,000-rpm range.

Gas turbines are of two types: single shaft and two shaft. The two-shaft type is generally used for pumping service because of the need for a variable load. Single-shaft turbines are more appropriate in constant speed service, such as driving electricity-generating equipment. In a two-shaft gas turbine, the gas-producer section is on a separate shaft from the power-producing turbine, permitting variation in the power turbine speed. In turn, by varying the pump speed, flow capacity can be controlled, making this type unit appropriate for pipeline pumping service. Speed can typically be adjusted from 70% to 110% of normal full-load speed.[7]

Liquid fuel or natural gas provides energy to the gas-producer section of the turbine. The atomized fuel or gas is mixed with compressed air and is ignited in this section. The resulting hot exhaust gases turn the power turbine shaft.

Gas turbines can also be classified as simple-cycle or recuperated turbines. In recuperated turbines, the heat of the exhaust gases resulting from burning fuel in the turbine is used to increase the turbine's efficiency.

Gas turbine manufacturers can provide fuel systems that allow a turbine to operate on fuels ranging from diesel through LP-gas in addition to natural gas. In one large crude pipeline system, gas turbines that power main line pumps are designed to be operated on three fuels at any given time.[8] Diesel fuel is used for starting, natural gas liquids are the primary fuel, and crude oil can be used as an emergency fuel. In this installation, a test was run to determine whether the natural gas liquids should be introduced into the turbine as a liquid or be atomized. It would have been possible to burn the NGL directly, but expected higher maintenance costs and changes in the properties of the NGL indicated vaporized NGL would provide the most stable system. Tests were also made using crude oil as fuel to determine the best choice of nozzles, and turbine blade and vane coatings for this service.

Such multifuel systems add cost to a gas turbine installation. The system is also much more complex, requiring sophisticated control systems and associated equipment to filter the turbine fuel and to transfer operation from one fuel to another. Shut-off valves, fuel transfer valves, and dual-fuel nozzles are also needed.

Gas turbine manufacturers provide data on the operating characteristics of a turbine, typically based on pressure at sea level and 59°F. Data determined on this basis provide what is known as an ISO rating and include the following:[9]

1. Power output vs speed at inlet temperatures.
2. Fuel consumption vs speed at inlet temperatures.
3. Fuel consumption vs output power and exhaust heat flow at inlet temperatures.
4. Horsepower correction vs altitude above sea level.
5. Horsepower correction vs inlet/exhaust pressure losses.

Operation. Starting a gas turbine is relatively complex. Compressed air, natural gas turbines, and AC and DC motors can be used for starting. Starting is normally controlled by a system that performs pressurizing and purging cycles. When various safety barriers are satisfied, the turbine is accelerated to about 40% of full-load speed. Fuel is then supplied to the turbine. When ignition occurs, the turbine accelerates to full-load speed. The control system then takes over, monitors turbine speed, and makes adjustments to meet station operating conditions.

Gas turbines operate at various speeds, depending on the type and size. The lower-speed industrial turbine may operate at a speed compatible with the pump or compressor being driven and may be used in a direct-drive configuration. But the high-speed aircraft-derivative turbines typically require speed reducers between the turbine and the pump or compressor.

Gas turbines can be the best approach to many pipeline pumping and compression services. Two-shaft turbines have a speed range that can meet varying flow requirements and, depending on the operating speed of the turbine, direct connection without speed reducers is possible. Other characteristics of the turbine, however, must also be considered in comparing prime mover alternatives. Fuel consumption at rated load is relatively high, and the turbine generally has a poor efficiency—and high fuel consumption—at partial loads.

Ambient temperature has a significant effect on turbine capacity. At low ambient temperature, the turbine can be substantially overloaded relative to its rated load at base temperature. This can be an advantage in the case of gas compression because peak demands on the pipeline system usually occur during periods of low ambient temperature. At higher temperatures, however, load capacity of the turbine is reduced below its rated load at base conditions.

Economics.[10] The initial cost of gas turbine prime movers ranges from $150–$200/bhp for the aircraft-derivative type, including a supervisory control system and starting equipment. The slower-speed heavy industrial gas turbines typically cost $170–$250/bhp, including control system and starting equipment. Special options such as multifuel systems and special control systems increase these costs.

Maintenance cost for the slower-speed industrial gas turbine is less than the maintenance cost for the aircraft-derivative unit. The range is $5–$10/bhp/year for the industrial unit, compared with $20–$30/bhp/year for the aircraft type. The heavier industrial, slower-speed turbines are less costly to maintain, in part because the time between overhauls is generally considerably higher than is the case with aircraft-type units. One estimate indicates the time between overhauls for an aircraft-type turbine is about 25,000 hr; for the industrial turbine, it can be 100,000 hr. A major inspection of the aircraft turbine is needed about each 6,000 hr, while an industrial turbine may operate up to 30,000 hr without a major inspection.

As is the case with any prime mover, availability and cost of energy (fuel) is an important consideration in evaluating alternatives. Since gas turbines can operate on a variety of fuels, there is a flexibility not available with other prime movers. But choosing the type of fuel to use can be difficult because of the changes in cost over the life of the project. At the time a prime mover is selected, the relative cost of natural gas and natural gas liquids, for instance, may be significantly different than they are after the unit has been in operation for some years. The cost of each type of fuel should therefore be estimated. It is practical, however, to change the type of fuel used if it is expected that a different fuel will be less costly for a long enough period to warrant the cost of converting the fuel system.

Fuel consumption can be expressed in BTU/bhp-hr in the case of natural gas or in lb/bhp-hr or other units for liquid fuels. In one comparison, an aircraft-type turbine is estimated to use about 0.51 lb/bhp-hr of No. 2 diesel or 9,000–10,000 BTU/bhp-hr when fueled with natural gas. A heavy industrial turbine typically uses 0.59 lb/bhp-hr of diesel and 10,000–12,000 BTU/bhp-hr when fueled with natural gas.

Reciprocating engines

Some diesel engines are used for pumping, but the most common pipeline application of reciprocating engines is in gas compression service. Reciprocating engines and compressors and integral engine/compressor units are available in large sizes; rated horsepower can range up to several thousand horsepower for a single unit.

Integral compressor/engine units combine both engine and compressor on a single frame. The crankshaft serves both the engine's power cylinders and the cylinders of the compressor. In large integral units, power cylinders are typically vertical or are arranged in a vertical V design. Compressor cylinders are horizontal. Both power and compressor "sides" of the integral engine/compressor have pistons, connecting rods, exhaust and intake valves, rod bearings, and other similar components. Natural gas is a common fuel for these units. The reciprocating engine/compressor has been the workhorse of natural gas transmission and gathering for many years. It is a dependable design.

Many of the reciprocating engine/compressor units now in natural gas pipeline service were installed when fuel costs were very low. One report indicates most compression equipment installed between 1955 and 1965 in North America was of the reciprocating engine/compressor type.[11] Much of this equipment is still in good operating condition, but recent efforts have concentrated on improving the efficiency of these compressors. Fuel consumption of reciprocating engine/compressors can be reduced from 5–15%. Even the efficiency of units built in recent years can be improved.

The efficiency of the engine side depends on several variables, including air-fuel ratio, ignition timing, maximum torque, and the number of engines required to compress the desired volume of gas. Air-fuel ratio affects combustion efficiency; efficiency decreases when the mixture is either too lean or too rich. This ratio is controlled by varying the flow of air to the engine cylinders. Combustion efficiency also depends on the point in the combustion cycle at which ignition occurs. This point is changed by adjusting the timing. Torque is changed by varying the piston displacement of the compressor cylinders. Losses in reciprocating engine/compressor units result from driving auxilliary equipment and friction and typically amount to 10–15% of the unit's total horsepower output.

Reciprocating engines are available in either naturally aspirated or turbocharged models. In naturally aspirated engines, the exhaust gases are forced out by the pressure created by the piston. The amount of air drawn into the cylinder is limited by exhaust gases remaining in the clearance volume of the cylinder, the volume at the end of the cylinder that is not "swept" by the piston. Turbocharged, or supercharged, engines are equipped with a mechanically driven blower or exhaust-driven turbocharger that supplies air under pressure to the cylinder. This increases the amount of air entering the cylinder and helps displace exhaust gases.

The horsepower rating of these engines must be lowered when used at altitudes greater than 1,500–2,500 ft above sea level: The altitude above which the engine is derated depends on the individual manufacturer. Beyond the altitude at which derating begins, horsepower rating is typically decreased by about 3% for each additional 1,000 ft of altitude.

Some engines are also derated for ambient temperatures above a certain temperature, but this is not as common in reciprocating engines as altitude derating.

Lowering the horsepower rating of an engine is done to reflect a more accurate horsepower output of a given engine under specific conditions of ambient pressure and temperature. In the naturally aspirated engine, for example, atmospheric pressure forces air into the cylinder. If atmospheric pressure is lowered—it decreases with increasing altitude—less air will be drawn into the cylinder and the engine will not develop as much horsepower as it would at a lower altitude.

To operate reciprocating engines most efficiently, the engine cylinders must be balanced. Balancing adjusts the parameters affecting the combustion cycle so each engine cylinder is accepting an equal share of the total load on the engine. Methods for determining if engine cylinders are balanced vary; comparing exhaust temperature was a traditional way. Most techniques now used are based on measuring either peak pressure or average pressure in the cylinder.

Engine performance is also evaluated based on mean effective pressure (mep) in the cylinder. During the combustion cycle, pressure in the cylinder varies according to the point in the cycle. Mean effective pressure is the constant pressure acting through the stroke that would produce the same work as the variable pressures during the cycle. Mean effective pressure is used, along with the piston stroke length, piston area, and number of power strokes per minute to calculate an indicated horsepower (ihp) for each cylinder. Adjustments are then made so that the ihp's of each cylinder are equal.

Auxiliary services for reciprocating compressor engines are important to efficient operation and preventing downtime. Lubricating systems can be quite complex for a large integral engine/compressor unit, for example, since both power-end and compressor-end cylinders require adequate lubrication to prevent excessive temperatures and damage. Cooling is also important to dissipate the heat generated by combustion in the power cylinders and by gas compression in the compressor cylinders. Both these systems—and other engine conditions—

are monitored by sophisticated systems that can shut the engine down if operating conditions exceed prescribed limits.

Reciprocating engines and compressors can cause more vibration than gas turbines or electric motors if improperly designed and installed. Concrete foundations are usually required for large reciprocating engine/compressor units to provide stability and absorb engine vibration. Piping connecting the compressor must also be designed and installed in a way that prevents vibration from being transmitted to other equipment in the station.

Economics. In general, reciprocating engines have a higher initial cost than other types of prime movers. Oil consumption is also higher than that of other types of drivers, but maintenance costs can be lower in many cases. With modern monitoring equipment, it is practical to run engines longer between overhauls. Performance data gathered by a monitoring system can now be used to schedule maintenance efficiently. One gas transmission company traditionally overhauled completely each of 184 reciprocating engines on an annual basis because it was the only way to determine the condition of the engine. But an engine analyzer and computer testing program has made it common to run those same engines for several years between overhauls.[12]

REFERENCES

1. Earl Seaton, "U.S. Pipelines Keep Energy Moving," *Oil & Gas Journal*, (22 November 1981), p. 73.
2. *Gas Engineers Handbook*, 1 Ed. New York: Industrial Press, 1974.
3. Ralph E. Wright, "Prime Mover Selection—1: Here's Help for Selecting Pipeline Prime Movers," *Oil & Gas Journal*, (16 February 1981), p. 87.
4. Ralph E. Wright, "Prime Mover Selection—2 (Conclusion): How Motors Compare With Turbines as Prime Movers for Pipeline Pumping," *Oil & Gas Journal*, (23 February 1981), p. 77.
5. See reference 3 above.
6. See reference 3 above.
7. See reference 4 above.
8. Chester Stasiowski, "Control System for Saudi Arabian Crude Line," *Oil & Gas Journal*, (28 December 1981), p. 201.
9. See reference 3 above.
10. See references 3 and 4 above.
11. Hans D. Lenz, "Microprocessors in Pipelining: Microprocessor-Based Optimizing Systems Boost Engine/Compressor Unit Efficiency," *Oil & Gas Journal*, (14 December 1981), p. 151.
12. James O. King and Neil Goodman, "Preventive Maintenance Keeps Compressor Engines at Peak Efficiency," *Oil & Gas Jounal*, (12 April 1982), p. 111.

7

CONSTRUCTION PRACTICES AND EQUIPMENT

PIPELINE construction methods differ depending on the geographical area, the terrain, the environment, the type of pipeline, and the restrictions and standards imposed by governments and regulatory agencies. However, construction techniques can be broadly classified as land, offshore, and Arctic. The biggest differences exist between land construction (including Arctic) and offshore construction.

As outlined in Chapter 1, construction costs also vary according to location, line size, environmental conditions, equipment required, and the construction schedule. Considering all types of pipeline construction, construction costs account for about 40% of the total investment in a pipeline system.[1]

In general, construction of a pipeline on land is the least expensive of the three types. Both offshore and Arctic pipeline construction are very expensive, but it is difficult to generalize about the relative cost of the two. The cost of installing a pipeline in the Arctic can be much greater than the cost of many offshore pipelines, but large pipelines built in difficult offshore environments are also very expensive.

Despite many differences, all pipeline construction projects have a number of features in common:

1. The methods of designing the system—arriving at the optimum pipe diameter, determining the amount of horsepower required for pumping or compression, meeting safety standards—are similar for all pipelines.
2. There are a number of design criteria that are set by government or regulatory agencies to insure safe operation of a pipeline and the safety of personnel and property near the pipeline. These standards vary depending

126

on the location of the pipeline, both geographically and in relation to populated areas and other facilities.

3. Most oil, gas, and products pipelines are constructed by welding short lengths, or *joints*, of pipe together. There are a few exceptions to the use of welded connections, but these are in short lines within a producing field or in similar applications.

4. Extensive testing of welders and the welds they produce is an important part of the construction of all long-distance petroleum pipelines.

5. Almost all oil and gas pipelines are buried below ground level; even most offshore pipelines are buried below the sea bed for protection. There are cases in which large segments of a major pipeline are not buried, the most notable example of which is the trans-Alaska crude pipeline where above ground sections were installed to protect permafrost areas.

6. All pipelines are tested for leaks following construction before the line is put in service. Several techniques can be used, but the most common is hydrostatic testing—filling the line with water and subjecting it to a pressure greater than the design operating pressure.

7. Most pipelines are coated on the exterior to prevent corrosion. Offshore pipelines are also "weight-coated" with a concrete coating to overcome the force of buoyancy and to prevent the pipe from floating to the surface.

8. Most pipelines must have one or more pumping stations or compressor stations along the route to provide energy to overcome pressure loss and keep the fluid in the pipeline moving.

9. The construction of all pipelines follows this general sequence: design and route selection, obtaining rights of way, installation, tie-in to origin and destination facilities and pumping or compressor stations, and testing.

Major pipeline projects are built by pipeline construction contractors rather than by the company that will own and operate the system. Typically, several contractors are invited to submit bids on the work. More than one contractor may be involved in a single large pipeline project to speed completion of the line. In addition, a number of other firms are involved to supply pipe and equipment and to perform special services such as pipe coating. Pump or compressor station construction may be handled by yet another contractor or group of contractors, depending on the size and complexity of the stations. If offices and control rooms are required, additional firms specializing in equipment or services related to those facilities may be involved.

On offshore pipeline construction projects, a number of miscellaneous services are required, including crew transportation and food service.

Land pipeline construction

Construction of a pipeline on land may involve relatively mature technology, but it is often a challenge to obtain rights of way and the required permits for construction and operation.

Proper route selection during the design stage can minimize some of these problems and reduce construction costs. The general route between the origin and destination of the pipeline is apparent when the pipeline project is conceived, but slight changes in the route to reduce the number of rivers and highways that must be crossed or to avoid abrupt changes in elevation can mean substantial reductions in construction cost. Some changes may also be dictated by features or terrain where pipeline construction is prohibited, such as designated wilderness areas.

The economic advantages of each change must be considered. If a steep slope can only be avoided, for instance, by increasing the length of the pipeline significantly, it may be more economical to traverse the slope. Changes in the route can affect pipeline operating conditions to the extent that additional pump or compression horsepower may be required. The additional cost of pumping or compression over the life of the system must be compared to the savings possible by a change in route.

On a large project, the first step in the field may be to obtain aerial photographs of the proposed route. The route is plotted on these photos and a survey is begun. As the route is fixed and surveying proceeds, acquisition of right of way can start. Each landowner's tract of land is plotted on an alignment sheet of the pipeline route, and each tract's position can be accurately described in relation to known points in an existing survey system.

Elevations along the pipeline route relative to a datum must also be obtained in order to plot a profile of the proposed line. These elevations are needed for completing the final design, since the change in elevation along the route is needed to calculate flow capacity, amount of pump or compressor horsepower required, and spacing of pump or compressor stations. Elevation data may also indicate the need for additional equipment or changes in operating conditions if steep slopes or abrupt changes in elevation are encountered.

The portion of the plan on which survey information is recorded typically has the pipeline's origin as a horizontal datum. Then mileposts designated along the line can be used as reference points for locating other facilities and recording the location of work or equipment. Some vertical datum is chosen, and elevations are then indicated at intervals along the route in feet above or below that datum. The result is that any point on the pipeline can be located both vertically and horizontally in relation to a point of known elevation above sea level and in relation to a horizontal point in the area's existing survey system.

After the route has been determined, right of way must be obtained throughout the length of the route before construction can begin. The right of way is the area alongside the pipeline where construction operations are performed. It also permits the pipeline owner to have access to the pipeline at any point in case repairs or maintenance are needed during operation.

Width of the right of way varies according to the size line, the type of terrain, the construction method to be used, and any special restrictions. Typically, the right of way width for a large-diameter, long-distance crude or natural gas pipeline is 50 ft. An example right of way schematic is shown in Fig. 7–1. In that project, aboveground vegetation was cleared from a portion of the right of way averaging about 35 ft wide for operation of construction equipment, but only the portion of the right of way needed for construction was cleared. In some areas of rough terrain, a 50-ft right of way clearance was required. In this example project, expanded areas of right of way were required at major river crossings. Two areas of 250 ft by 450 ft were needed for storage and equipment, one for each side of the crossing. Smaller areas were needed at each side of road, railroad, and minor river crossings. Of the total 50-ft right of way, 35 ft was used for working space and 15 ft was used for ditch spoil.

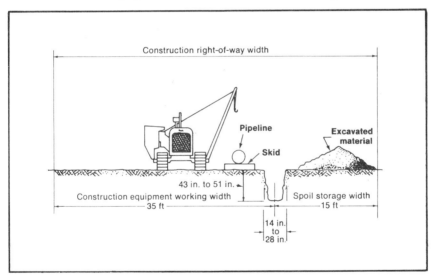

Fig. 7–1. Pipeline construction right of way. Source: *Oil & Gas Journal,* 16 June 1980, p. 96.

At points along the pipeline where pumping or compression facilities must be built, a larger area must be obtained. The size of these areas depends on the size of the facilities to be installed. It is desirable to obtain a right of way as narrow as will be practical during construction because obtaining right of way is a complex procedure involving many individual landowners. Many resist the idea of a pipeline traversing their property and value highly the right of way they are asked to sell to the pipeline company.

Still, a right of way that is too narrow and restricts the operation of construction equipment can slow construction and increase costs. The result is that pipeline companies try to buy the right of way they need—no more.

Interstate gas pipelines in the United States also have the right of eminent domain, which allows the government or a public utility (a common carrier in the case of a pipeline) to take private land for public use through condemnation proceedings.

During construction, all equipment and the pipeline ditch must be contained within the right of way or additional damage payments will be due the property owner. Equipment involved includes bulldozers used for clearing the route and for lowering the pipe into the ditch, ditching machines, welding equipment, coating and wrapping equipment and weld-inspection equipment. Pipe must also be delivered to the job site along this right of way, usually by truck.

Construction can take place within a relatively narrow right of way because pipeline construction equipment is distributed along the pipeline (Fig. 7–2) route in a sort of "moving assembly line." Only one major item of construction equipment is normally needed at any one point along the line. The distance along the pipeline over which this equipment is deployed is relatively short,

Fig. 7–2. Construction crews lay pipeline. Source: *Oil & Gas Journal,* 15 October 1979, p. 117.

typically less than a mile, but there may be several sets of construction equipment along the pipeline route. These complete sets of equipment—for ditching, welding, coating, lowering in, and backfilling—are called *spreads*. A single pipeline may be built using several spreads, reducing construction time by severalfold.

A very large pipeline project may even be divided into two or more segments, and a different construction contractor will install each segment. Each of these contractors may field several spreads to build his segment of the pipeline.

Obtaining the required permits, especially in the United States where environmental and pollution laws are strict, can be one of the most time-consuming steps in pipeline construction. Permits are needed for crossing roads and streams, for passing through federally owned lands, for traversing designated wildlife feeding and breeding areas, and for crossing wilderness areas, for example. An environmental impact statement (EIS) must also be prepared for most pipeline projects in the United States. The EIS not only must deal with environmental conditions along the main pipeline, but it also must meet restrictions on air emissions and other discharges associated with pumping and compression stations.

The trans-Alaska crude pipeline is an example of complex permitting on a large pipeline project. In addition to wildlife areas and emission restrictions, special requirements to protect permafrost areas had to be met with innovative design and construction techniques. Other examples of complex permitting procedures were the proposed Northern Tier pipeline and the Pactex pipeline.

The Northern Tier crude pipeline, proposed in 1974, was to be a 1,500-mile, 40-in. and 42-in. diameter pipeline from Port Angeles, Washington, to Clearbrook, Minnesota. By mid-1981, the company proposing to build the project had spent $50 million on the project, and construction had not yet begun. At that time, a judge, after 18 months of hearings on the proposal, recommended that the application be denied because the application was "inadequate."[2] In early 1981, the permitting process had been underway for five years; by mid-1982, about 1,400 permits had been obtained from federal and state governments. At that time, the proposal was modified in an effort to overcome objections to the proposed route and design. But in early 1983, the project was finally cancelled after about $60 million was spent.

Permitting delays and other red tape contributed heavily to the abandonment of the Pactex pipeline, a proposed crude pipeline to carry tankered Alaskan crude from Long Beach, California, to Midland, Texas, and points beyond.[3] The company proposing the pipeline abandoned the project after four years. Amounts spent on this project, including cancellation charges, amounted to about $50 million. An example of problems faced in this case was an overlap of regulatory involvement by two state agencies concerning air pollution caused by tanker unloadings at Long Beach. A spokesman for the firm planning the 500,000 b/d pipeline and associated terminals said the project bogged down in "a quagmire of federal and state regulations."

A further example of constraints that can be put on pipeline construction is shown in Table 7–1.

Installation.   The actual installation of the pipe for a pipeline includes these major steps:

- Clearing the right of way as needed
- Ditching
- Stringing pipe joints along the right of way
- Welding the pipe joints together
- Applying coating and wrapping to the exterior of the pipe (all except a portion of the pipe at each end is sometimes coated before being delivered to the job site)
- Lowering the pipeline into the ditch
- Backfilling the ditch
- Testing the line for leaks
- Cleanup and drying the pipeline after testing to prepare it for operation

Clearing of right of way is done first, using bulldozers or similar blade equipment. The amount of clearing required varies widely. Sometimes only one "pass" down the right of way with the bulldozer is required. Where the route passes through rough or forested terrain, however, clearing can be much more extensive. The purpose is to make it possible to move construction equipment along the right of way as needed.

The ditch, or trench, in which the pipeline will be installed is usually made to one side of the center of the right of way rather than in the center to provide adequate room for construction equipment and operations alongside the pipe. Typically, dirt excavated from the pipeline trench is deposited on the side of the ditch closest to the edge of the right of way; construction operations are conducted on the other side of the ditch. Ditching in relatively soft soil is done by a machine with a large wheel on which cutting teeth are mounted. The wheel rotates continuously as the machine moves along the pipeline route, and excavated material is continuously deposited alongside the ditch. In loose rock or hard soil, it may be necessary to use other equipment for trenching—a backhoe or clamshell bucket, for example. Blasting can be required when the ditch must pass through solid rock.

The depth of the ditch is based on specified *minimum cover*, or distance from the top of the buried pipe to the ground surface. For the same minimum cover requirements, a larger-diameter pipe requires a deeper ditch. The minimum cover varies according to requirements of regulatory agencies, the type of area

TABLE 7-1
Construction Date Constraints

Approximate mile posts	Dates construction will be avoided	Reason
MAINLINE		
65 to 79	Apr. 1 to May 31	Lesser prairie chicken booming period
418 to 420	Apr. 1 to May 31**	Big game and fish—important habitat
424 to 443	Dec. 1 to Apr. 15**	Big game crucial winter range
459 to 467	Nov. 15 to Apr. 30#	Big game crucial winter range
564 to 566	Nov. 1 to May 15§	Big game crucial winter range
584 (Kane Springs Canyon)	Mar. 15 to June 15§	Crucial riparian habitat
601 to 605	Mar. 15 to June 15§	Crucial wetland habitat
604 (Colorado River)	July 1 to July 31§	Crucial fish spawning
664 to 677	May 15 to June 20§	Pronghorn antelope fawning area
694 to 700	Mar. 1 to Apr. 30#	Chukar breeding grounds
697 to 705	Nov. 15 to Apr. 30#	Big game crucial winter range
785	July 1 to July 31§	Crucial fish habitat
788 to 800	Nov. 15 to Apr. 1‡	Big game crucial winter range
800 to 804	Mar. 15 to June 1‡	Sage Grouse strutting grounds
812 to 820	Dec. 15 to Apr. 1†	Big game crucial winter range
835 to 843	Dec. 15 to Apr. 1†	Big game crucial winter range
850 to 852	Mar. 1 to June 15*	Sage Grouse strutting grounds
858 to 862	Mar. 1 to June 15*	Sage Grouse strutting grounds
GATHERING LINES		
East line		
2 to 3	Mar. 15 to July 1†	Raptor nesting area
2 to 5	Mar. 1 to June 15*	Sage Grouse strutting grounds
18 to 19	Mar. 15 to July 1†	Raptor nesting area
17 to 20	Dec. 15 to Apr. 1†	Big game crucial winter range
74 to 77	Dec. 15 to Apr. 1†	Big game crucial winter range
West line		
17 to 18	Mar. 15 to July 1†	Raptor nesting area
24 to 44	Dec. 15 to Apr. 1†	Big game crucial winter range
84 to 96	Oct. 15 to May 15*	Big game crucial winter range
114 to 122		
(perennial streams)	May 15 to Aug. 15‖	Cutthroat trout spawning area
North line		
5 to 9	Oct. 15 to May 15*	Big game crucial winter range
46 to 59	Oct. 15 to May 15*	Big game crucial winter range
Northwest line		
2 to 5	Oct. 15 to May 15*	Big game crucial winter range

*Harrison, K. E., U.S. Bureau of Land Management, Kemmerer, Wyoming.
†Haverly, S. J., U.S. Bureau of Land Management, Rock Springs, Wyoming.
‡Smith, D. A., Utah Division of Wildlife Resources, Vernal, Utah.
§Wilson, L. J., Utah Division of Wildlife Resources, Price, Utah.
‖ Rensel, J. A., Utah Division of Wildlife Resources, Ogden, Utah.
#Whitaker, A. Colorado Division of Wildlife, Denver, Colorado.
**Gates, J., New Mexico Department of Game and Fish, Santa Fe, N.M.

Source: Seaton, *Oil & Gas Journal,* 16 June 1980, p. 96

through which the pipeline passes, and features along the pipeline route. A minimum of 3 ft of cover is typical, but it may be less in open, unpopulated areas and more when the pipe passes under roads, rivers, and highway borrow ditches. One project, for instance, specified a 4-ft cover under highway borrow ditches. Minimum cover (between the river bed and the top of the pipeline) was also set at 4 ft when crossing under rivers.

Width of the pipeline ditch varies according to the size of the pipeline. Typically, this width ranges from 14 in. to 28 in. for the intermediate pipeline diameters.

Normally, pipe is delivered to the right of way on trucks and is placed along the pipeline route so that each joint only needs to be moved over to the ditch when it is ready to be welded into the pipeline. *Stringing* the pipe joints along the right of way in this manner speeds construction. In some cases, pipe is brought to the job site in sections consisting of two single joints of pipe. This "double-jointing" saves welding time on the job site, which often must be done under less desirable conditions than exist in a fabrication yard.

With the ditch made and the pipe delivered, welding can begin. The pipe joints are placed over the ditch for welding. As welding proceeds, a section of pipeline steadily increasing in length is in place above the ditch. Pipeline welding is done with electric welding equipment, both manual and automatic (see Chapter 8). On a land pipeline construction job, welding machines are typically mounted on small trucks or pickups. The machines may also be mounted on tracked vehicles. A number of welders—each with a welding machine—work on each pipeline spread. Since a number of weld passes (a "bead" of weld material around the circumference of the pipe) must be made at each joint, a typical procedure is to have one welder make the initial passes at each joint. Other welders follow behind this lead welder, building up the weld metal at the joint by making additional weld passes until the required number of passes have been deposited. The number of weld passes required depends on the wall thickness of the pipe and its physical characteristics and is specified in the construction plans. The initial weld pass is one of the most critical.

It is important that the two ends of pipe to be welded are properly aligned so the weld will be uniform around the circumference of the pipe. Line-up clamps are used for this purpose at each joint before welding begins. After the first passes have been made, the alignment clamps can be moved to the next welding station.

A key part of pipeline welding is inspection. For most projects, welders must be qualified by testing on the size and type of pipe to be used on the job. After the welds on the pipeline are made, however, they must be thoroughly examined to insure the safety of the pipeline. The use of radiographic, or X-ray, examination of completed welds is the most common inspection method.

Construction plans specify what type of inspection will be required and what portion of the welds must be examined by each method. For instance, it might be specified that where the pipeline traverses open areas, 10% of the welds must be X-rayed; where the pipeline passes under railroads, highways, or rivers, all welds must be examined using radiography. In the X-ray inspection process, the film wrapped around the circumference of the pipe over the weld is exposed to radiation. When the film is developed, bubbles, cracks, slag inclusions, and other defects are visible. It is desirable to make this inspection and find any defective welds before the pipe is buried because defective welds must be removed and a new weld made.

As welding proceeds along the pipeline, a slight change in direction or a significant change in elevation may require a bend in the pipeline. Many such bends are made by a bending machine (Fig. 7-3) on the job site that bends a joint of pipe to the required curvature. Even large-diameter pipe can be accommodated in today's modern bending machines, but it may also be necessary to make some bends in a shop on a special machine. Depending on the diameter and the wall thickness of the pipe, slight changes in elevation may be accommodated by flexing the pipe without machine bending. Very small changes in direction may sometimes be made by letting the pipe lie to one side of the ditch or the other. But changes in direction or elevation without bending must be small, especially when large-diameter, heavy-wall pipe is being used.

Fig. 7-3. Pipe is bent with field-bending machine. Source: *Oil & Gas Journal*, 20 November 1978, p. 110.

In all bends, care must be taken to avoid deforming the pipe in the bend section. A bend with too short a radius can buckle the inside of the bend, reducing the strength of the pipe.

After welding is complete, the pipe exterior is "coated and wrapped." Coating and wrapping is done using special machines that move along the pipeline right of way. Coal tar enamel is the most common pipeline coating; others include thin-film powdered epoxy and extruded polyethylene.[4] Asphalt enamel and asphalt mastic are also used as pipe coating materials. Tape is then wrapped over this coating to provide additional protection to the pipe and to protect the corrosion coating, especially through rocky areas that might damage the pipe coating.

In some cases, coating and wrapping is "yard applied" to the pipe before the pipe is delivered to the job site. When this is done, a short distance at each end of the pipe joint is left bare to permit welding. Then these areas are coated and wrapped over the ditch after welding is complete.

Up to this point, the pipe is suspended over the ditch. Individual joints have been welded together, forming a continuous pipeline, and coating and wrapping is complete. The pipeline is suspended over the ditch by *sideboom tractors,* crawler tractors with a special hoisting frame attached to one side. Now the pipeline is gradually lowered to the bottom of the ditch ("lowering in"). It is sometimes necessary in rocky soil or solid rock to put a bed of fine soil in the bottom of the ditch before lowering in the pipeline. The fine fill material protects the pipe coating from damage.

Backfilling of the ditch on top of the pipeline is then done with the soil excavated from the ditch. Backfilling may be accomplished with a machine that uses an auger (Fig. 7–4) to move soil continuously from beside the ditch into the ditch as it moves along the pipeline. Other blade-type equipment may also be used to backfill the ditch. The elevation of the ground above the pipe after backfilling may be higher than the natural ground surface to allow for settlement.

Road/river crossings. Even small pipeline projects often involve crossing roadways and streams; a long-distance pipeline may cross scores of each. A variety of techniques is used for crossing these obstacles, depending on the length of the crossing, the size of the stream or roadway, and regulations. Crossing roadways can be done by either ditching or boring. When ditched, the roadway must be repaired, so this method is often not permitted. Boring is done with a horizontal boring machine that drills a hole under the roadway without disturbing the road surface. A conductor pipe is normally installed in the bored hole; then the pipeline is placed inside the conductor. Spacers are used to center the pipeline within the conductor to reduce corrosion.

Fig. 7–4. Pipe bedding material is placed in ditch. Source: *Oil & Gas Journal,* 14 August 1978, p. 63.

A backhoe or dragline can be used on minor stream crossings to make a ditch for the pipe to rest in. The ditch is then backfilled, and the pipe may be fitted with concrete weights to hold it in place against the stream currents and movement of the stream bed sediments (Fig. 7–5).

A technique for crossing larger bodies of water that has gained favor in recent years is directional drilling. It offers several advantages, including no disruption of traffic on the waterway and minimum environmental impact. Directional drilling has long been used in drilling oil and gas wells, but use for pipeline crossings was developed only a few years ago. One of the longest crossings using directional drilling was a crossing for a 22-in. diameter pipeline under the Orinoco River in Venezuela, a distance of 4,550 ft.[5] On this project, the drilling rig was positioned so the bit entered the ground at about a 12-degree angle. A pilot hole was drilled and then reamed to the desired size, and the pipe was pulled through the curved hole.

An instrument package behind the downhole drilling motor measures the direction and inclination of the bit and transmits the information to the operator. He can then change the drilling conditions to maintain the prescribed drilling path. In this project in Venezuela, the drill exited on the other side of the river within 12 meters of the target.

Fig. 7–5.   Pipe being installed in river crossing. Source: *Oil & Gas Journal,* 15 October 1979, p. 117.

Not all streams are crossed by installing the pipeline beneath the stream. Some pipelines are installed on "pipeline bridges," steel structures built to suspend the pipeline above the stream. Use of this method depends on a number of factors, including the presence of traffic on the waterway. The size and complexity of these pipeline bridges vary widely; one of the most complex examples is the pipeline suspension bridge built across the Tanana River in Alaska to carry the 48-in. trans-Alaska crude pipeline. With a clear span of about 1,200 ft, the bridge cost about $5 million and was one of 122 river and stream crossings required during construction of the line. Most of the other elevated crossings were made with girder bridges rather than suspension bridges.

In addition to supporting the pipeline, the Tanana River bridge had to meet special design considerations, including winds up to 100 mph, ambient temperatures ranging from −70°F to +100°F, extreme icing conditions, and seismic shocks of 7.5 on the Richter scale.[6]

Land pipelines must sometimes be anchored in place to prevent movement. Instances where anchoring is required include river crossings where currents can cause pipe movement or scour beneath pipe installed on the river bed, a dry wash subject to temporary flooding, or where certain types of backfill are used.[7]

All pipeline construction includes testing the completed pipeline before it is put into operation. On long pipelines, the line will normally be tested in sections; on short lines, the entire pipeline may be tested as a unit. A common approach is hydrostatic testing—filling a closed pipeline section with water, then pressurizing the line to a specified pressure to check for leaks. Temporary connections for filling and draining the pipeline are used, and a pump is used to "pressure up" the line. The pressure is maintained on the line for a specified time. If pressure declines, a leak is indicated. It then is necessary to find the leak and determine the cause. If a weld or the pipe itself is faulty, it must be replaced or repaired.

The pressure to be reached and the time it must be maintained are specified in the construction plans. In the United States, the Department of Transportation and other agencies specify the test pressure to be used based on the pipeline's location, its function, its design operating pressure, and other factors. For example, hydrostatic test pressure may be specified as 125% of the maximum design operating pressure of the line. Other test pressure specifications are related to the minimum yield strength of the pipe.

After the pipeline has been tested, it is important that moisture and foreign materials be removed from the pipeline before it is put into operation. Such materials could damage pumping, compression, and other equipment if swept into that equipment when the pipeline is put in service. Water, sand, dirt, welding slag, and some even stranger materials have been removed from newly completed pipelines.

The most common method of cleaning a new pipeline is by *pigging,* in which a device with cups or brushes that contact the wall of the pipeline is forced through the pipeline, moving debris ahead of it. The driving force for the pig is usually clean water. Another method of cleaning a new pipeline was used recently in the North Sea. There, a gas pipeline was cleaned using batches of plastic fluids (gels) and pipeline scrapers (pigs) to remove loose rust, silt, welding rods, weld slag, and other debris from the pipeline.[8] The method removed 350 tons of debris from the 230-mile long, 36-in. diameter line. Debris was entrained in the gels, and the cleaning train of gel batches and pigs was driven by pumping inhibited fresh water into the line at about 70 bbl/min. The train was moved through the line at about 1 ft/sec in 17 days.

Operators of the pipeline report that flow tests run prior to and after cleaning showed the cleaning operation significantly reduced pipe roughness. The operators considered other cleaning methods, including the common practice of using multiple mechanical pigs and the use of a high-velocity water flush. But the amount of debris expected to be in the line posed a danger of sticking mechanical pigs, causing the line to be blocked.

It is often necessary to dry natural gas pipelines after hydrostatic testing to prevent the formation of hydrates when the line is put into service. Gas hydrates are complex chemical compounds formed when free water is available in the presence of hydrocarbon gases. They are crystalline and resemble ice or snow. If allowed to build up on the wall of natural gas pipelines, they reduce flow efficiency by increasing friction and reducing the effective diameter of the pipe. In addition to the presence of free water, other conditions affecting the formation of hydrates include low temperatures, high pressure, and the density of the gas mixture.

One way to dry a natural gas pipeline is to move methanol through the line to absorb the moisture. The methanol, a liquid, may be moved through the line in several batches separated by pipeline pigs. The energy to force the pigs and methanol through the line can be furnished by the gas source that the pipeline serves.

Another approach to drying is to use dry air that has been processed in an air-drying plant to remove moisture. It is fed into the pipeline from the air-drying plant, and special foam pigs separate the moist air from the dry air.[9] The line is dried in sections, and when all tie-ins are made the pigs are sent through the entire line with dry air.

Station construction normally proceeds concurrently with pipeline construction. This work is usually done by separate contractors from those handling the pipeline installation, except on very small projects. Large projects in which pump or compressor stations are complex and involve control rooms, offices, and similar facilities may involve more than one contractor for station construction.

When the line is completed, the main line must be tied into the origin and destination facilities it serves and into pump or compressor stations and other equipment along the pipeline. Much tie-in work involves piping connection; on large projects, this can be complex. Instrumentation and metering systems must also be connected during tie-in. Offshore, tie-in work usually involves installation of the pipeline riser to connect the end of the pipeline on the ocean floor to piping on the above-surface platform. More detail on riser installation is given later in this chapter.

Offshore pipeline construction

Many operations are common to both onshore and offshore pipeline construction. The key design differences are that installation stresses rather than operating stresses often control the design of offshore pipelines. Environmental forces are also more significant offshore.[10]

Similar to onshore pipeline construction, offshore pipeline construction work is usually awarded to installation contractors after an evaluation by the owner of bids submitted by a number of contractors. These contractors own and operate the offshore pipeline construction equipment. Work can be awarded on a lump sum basis or on a day-rate basis. A day-rate basis is appropriate when design or survey data are not complete and the contractor cannot determine accurately the full extent of the work prior to submitting a bid. If detailed information is available on which the contractors can base their bids, the lump sum is usually in the best interest of the project owner.

Several construction methods can be used for offshore pipeline construction, including the conventional lay barge method, the reel-barge method, the vertical lay method, and the tow method. Of these, the vertical lay method is still experimental. All require large sophisticated marine vessels.

The most common method is the use of a conventional lay barge. It is a versatile technique, applicable to most pipe sizes and most water depths. The modern pipelay barge is a floating platform on which operations similar to those involved in building an onshore pipeline are conducted. A typical lay barge is fitted with three to six welding stations, an inspection station where welds are examined, and one or two field-joint coating stations. Individual joints of pipe (or double joints, in some cases) are welded together onboard the barge, and the completed pipeline exits the rear of the lay barge (Fig. 7–6). Some lay barges are equipped for both automatic and manual welding; some are equipped for only manual welding.

A key component of the lay barge is the tensioning system. Tensioners are required to hold the weight of the completed pipeline behind the barge and allow pipe to move off the barge at the desired rate as each new joint is welded into the line. These tensioners have "grippers" that hold the pipe and let it move off the rear of the barge in a controlled manner (Fig. 7–7). The amount of force that must be applied by the tensioners to hold the pipe on the barge varies with pipe size and weight and water depth. The capacity of the tensioning system is a key criterion in rating the lay barge for a particular project.

Tensioner capacity covers a wide range. Most lay barges have from one to three tensioners, with total tensioner capacity between 50,000 and 200,000 lb. A few large lay barges have higher tensioner capacity. Most lay barges are also equipped with a winch for abandoning and recovering the pipeline when it is necessary to suspend pipelay operations. Capacity of this winch is typically 150,000–300,000 lb.

Fig. 7–6.   Offshore lay barge installs pipeline. Source: *Oil & Gas Journal*, 3 January 1977, p. 53.

Fig. 7–7.   Grippers on lay barge tensioner hold pipe. Source: *Oil & Gas Journal*, 10 January 1977, p. 91.

Another important part of the conventional lay barge is the *stinger*. The stinger is used to support the completed pipeline as it moves off the lay barge into the water. Stinger design varies; there are straight, curved, and articulated stingers. The design required for a specific project is determined by water depth, pipe size and weight, and other conditions. One reason for the variety of designs is to reduce the stress on the pipe as it moves off the barge into the water and down to the ocean floor in the S-shaped configuration. The *overbend* area, the upper curve of the S shape, and the *sagbend,* the lower curve, are critical areas of design and installation. Too great a curve can stress the pipe and cause damage during laying.

Varying lengths and types of stingers have been designed to ensure this damage does not occur. One approach to reducing the curvature of this overbend has been the use of lay barges with a sloping ramp at the rear of the barge where the pipe enters the water. In general, the curvature of the overbend depends on the length of the stinger, and the curvature of the sagbend depends on the tension being applied by the barge's tensioners.

In moderate water depths and calm weather, stingers up to 600 ft long may be used. In more severe environments, such as the North Sea, shorter, heavier stingers are used because stingers are susceptible to damage in severe weather. Pipeline construction must often be suspended during severe weather to avoid stinger damage. Conventional lay barges may suspend operations when seas reach 6–10 ft; larger semisubmersible barges, because they are more stable, may be able to continue operation in seas up to 15 ft.

On the barge, operations are quite similar to operations along an onshore pipeline construction spread. At several welding stations, welders join the individual lengths of pipe. After a joint is partially welded, the pipe is moved ahead for completion of the weld. Radiographic inspection is done at one station, and the uncoated area of the pipe on each side of the weld is field-coated at another station.

In offshore pipeline construction, it is common to apply coating in an onshore yard before the pipe is delivered to the lay barge. As in onshore construction, a section of bare pipe is left at each end of each joint to permit welding. Then that area is field-coated on the lay barge after welding is complete. Double jointing may also be used, in which two single lengths are welded together onshore and pipe is transported to the lay barge in double lengths.

In addition to coating required for corrosion protection, offshore pipelines are coated with a layer of concrete. Used primarily to provide "negative buoyancy" for the pipeline—weight needed to keep the pipe on the sea floor— concrete coating also must resist damage during the laying and trenching. While in service, it must resist damage from fishing gear and other hazards. An important function of concrete coating on offshore pipelines is to protect the anticorrosion coating from damage. Concrete coating is applied by a variety of methods outlined in Chapter 3. The concrete mix, its thickness, and its strength vary depending on the location of the project and its design specifications.

All of the stations on the lay barge—welding, inspection, coating—remain in the same position on the barge. The pipe moves through these stations as the lay barge proceeds along the pipeline route. Most lay barges are held in position and are moved along the pipeline route with a system of anchors and anchor winches. Anchors must be moved ahead periodically as the barge progresses along the route.

A few deepwater lay barges are equipped with dynamic positioning, which permits operation without the use of anchors. The system includes a sophisticated position-monitoring system that accurately and continuously feeds the barge's position into a computer that controls the barge's positioning thrusters. If the barge begins to move away from the desired position, commands are sent to the thruster system that cause a specified amount of thrust to be applied in the proper direction to maintain the barge on station. This type of station-keeping system is only applicable to deepwater operations, and it is not used routinely. One reason for its infrequent use, except where it is the only practical solution, is that the almost constant use of thrusters to maintain position requires large amounts of fuel, adding significantly to operating costs. But this equipment, in commercial use, shows the extent of the industry's deepwater pipelaying capability.

A variety of sizes of conventional lay barges is in use today, each with a particular application range of water depth and pipe size. Typically, a modern lay barge contains personnel living quarters, galley and mess hall, and other necessary support services in addition to equipment directly used for constructing the pipeline.

Some lay barges are similar to cargo barges, while others have a more specialized design. For example, the semisubmersible lay barge consists of several columns that support an abovewater platform on belowwater buoyancy units, or pontoons. Semisubmersible lay barges are designed for use in more severe environments, greater water depths, and when laying large-diameter pipelines. The greater stability of the semisubmersible lay barges makes them able to continue operations under more difficult sea and weather conditions.

Some lay vessels also perform other functions. A derrick/lay barge is equipped with a large crane in addition to pipe-laying equipment for doing offshore construction work, such as setting offshore platform jackets and other structures. Lay/bury barges are equipped with jetting equipment for burying offshore pipelines.

Modern lay barges have permitted the industry to lay pipe as large as 42 in. in diameter in water depths greater than 1,000 ft. But smaller pipelines have been laid in much deeper water. The most outstanding project in this regard is the construction of portions of the Trans-Mediterranean Pipeline, designed to carry natural gas from Algeria to Italy. Sections of the route involve crossing both the Sicilian Channel and the Strait of Messina.

Portions of the Trans-Mediterranean Pipeline system indicate the industry's offshore pipelaying capability. In 1980, three 20-in. pipelines were laid in water depths up to 1,968 ft spanning a 99-mile route between Cape Bon, Tunisia, and Mazar del Vallo, Sicily. The entire system extends from gas fields in Algeria to Italy's Po Valley. The $3 billion system, over 1,500 miles long, will move 1.28 billion cu ft/day of natural gas.[11] The pipelay vessel *Castoro Sei*, owned by Saipem, was used for the deepwater sections of the job (Fig. 7–8). It is a twin-hulled, self-propelled, semisubmersible lay vessel capable of laying pipelines in water depths to 2,500 ft. Operating capabilities include a maximum significant wave height of 18 ft, wind speed of 50 knots, and a current of 2 knots. Survival conditions for the vessel are a maximum wave height of 80 ft and a maximum wind velocity of 100 knots (concurrent). The 498-ft-long vessel has eight welding stations, one field-joint coating station, one inspection station, and one X-ray control station. *Castoro Sei* is one of the most sophisticated lay barges available.

Reel barge. The reel-type lay barge was designed to lower offshore pipeline construction costs by increasing the laying rate and decreasing the time

Fig. 7–8. Semisubmersible pipe lay barge is designed for deep water. Source: *Oil & Gas Journal*, 30 April 1979, p. 158.

construction operations are exposed to offshore weather. It is used primarily for smaller line sizes—2-in. to 12-in. diameter—but the capability exists to lay pipe as large as 16 in. in diameter using the reel method.

The reel-type lay barge contains a large-diameter (40–60 ft) reel. Joints of pipe are welded together at an inland site and are wound on the reel. The barge with the full reel is then moved to the pipeline construction site. One end of the pipeline is anchored, typically at an offshore production platform, and then the barge moves along the pipeline route, unreeling the pipe. If the length of the pipeline is such that not enough pipe can be wound on one reel, additional reels can be used. The ends of the two reels must be joined at the construction site.

Since the yield point of the steel is exceeded when the pipe is wound on the reel—the pipe is bent and will not return to its original straight configuration by itself—straightening rollers are used as the pipe is unreeled into the water. Firms supplying this service have done considerable testing to prove no damage to the pipeline results from this bending and straightening cycle.

An example of a modern reel-type lay barge is the Santa Fe *Apache*. The *Apache* is capable of laying pipe as large as 16 in. in diameter by the reel method. The vessel contains a vertical reel on which pipe is spooled; the reel's capacity varies from about 50 miles of 4-in. diameter pipe to about 5.7 miles of 16-in. diameter pipe.[12] The 404-ft-long vessel can lay pipe at speeds up to 2 knots and is equipped with a dynamic positioning system that allows the vessel to remain in position without the use of anchors. *Apache* is also outfitted with a saturation diving system capable of operating in water depths to 1,500 ft.

The Santa Fe *Apache*'s first job was to lay four flow lines and two control umbilicals in the North Sea. In this project, 27 "stalks" of pipe were welded together at an onshore base for mounting on the pipelay reel. Each stalk was 900 ft long and contained 24 double random lengths of 4.5-in. OD line pipe coated with thin film epoxy. Completed stalks were moved from the welding area to pipe racks before being reeled onto the ship's reel. The stalks were joined as the pipe was fed up a ramp and through tensioners onto the pipelay reel.

Use of reel pipelaying has special application, currently to jobs involving smaller-diameter lines of moderate length. Where it is applicable, reel pipelaying can offer savings in construction costs by doing the welding at onshore sites and will probably become more widely used as the technique is developed.

Flexible pipe has also been installed offshore using a technique in which pipe is coiled in cylindrical baskets on the lay vessel. Pipe is then unspooled from the baskets as the vessel moves along the pipeline route. The first operation of this kind was in the North Sea in 1976 where a flow line was laid to connect a subsea well with production facilities.[13] The technique is particularly applicable to installing smaller-diameter, moderate-length offshore pipelines, such as field flow lines.

Vertical lay method. To overcome the problems related to excessive weights and stresses on the overbend when using a conventional lay barge equipped with a stinger in very deep water, a method has been proposed in which the pipe is welded together in a vertical position on the lay barge and is allowed to enter the water vertically below the vessel. A structure much like a derrick on an offshore drilling rig would support the pipe on the barge (see Chapter 12).

Pull methods. Another approach to offshore pipeline construction is the pull or tow method. No lay barge is used in this method; long sections of pipe are welded together onshore and pulled into the water a section at a time. The method is particularly applicable to crossing narrow, deepwater channels, but it can also be used for moderate-length offshore pipelines. The sections are pulled by winches in the case of a narrow crossing or by a tow vessel in the case of an offshore line.

This type of offshore pipeline construction has several advantages: much less expensive offshore construction equipment is required; much of the work is done onshore using conventional techniques; and the time the operation is exposed to severe offshore weather is reduced. Large-diameter lines (30-in.) up to 20 miles long have been installed using this method. Typically, though, projects using this technique involve 10–15 miles of 16-in. diameter to 24-in. diameter pipeline.

The tow method has been recommended in a number of applications:[14]

1. Near shore in shallow water where lay barge operation is not possible
2. For bundles of several pipelines or very large-diameter lines that are difficult to handle by lay barge
3. Where difficult or dangerous maneuvering by lay barge is required
4. In deep water, where the capacity of lay barge tensioners, stinger or positioning system is exceeded
5. In the Arctic, where heavy ice cover exists
6. Where only a short installation season is available because of high sea states or other environmental conditions

Four tow methods have been described. They include the bottom tow, the off-bottom tow, the near surface tow, and the constant depth tow (Fig. 7–9). In the bottom tow method, the pipeline—or bundle of pipelines—is towed along the ocean floor by a tug. The ocean floor where the pipeline is towed should be relatively flat and free of obstacles. In the off-bottom tow, the pipeline is floated at a distance off the ocean floor by adjusting buoyancy with weights and floats. When the pipeline is in position, the floats are either released to surface or flooded, allowing the pipeline to sink to the sea bed. There is less chance of damage to the pipeline with this method than when towing along the sea bed.

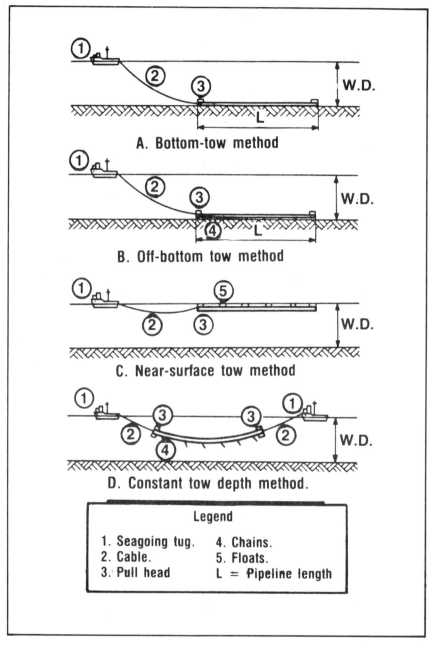

Fig. 7–9.   Tow methods for offshore pipeline construction. Source: *Oil & Gas Journal*, 22 June 1981, p. 64.

In the surface tow method, the pipeline is towed to its site while buoyed near the water's surface with floats or pontoons. Lowering of the pipeline in shallow water can be done by releasing the floats or pontoons in one step. In deep water, floats can be released successively to allow the pipeline to settle to the bottom in an S-curve configuration. Environmental conditions—wind velocity and wave height—have a significant effect on this technique while the pipeline is in the floating position.

Constant tow depth techniques offer the advantages of the off-bottom tow method—the operation is not significantly affected by seafloor conditions—and smaller tow forces are needed than those required for the bottom tow method.

All of these techniques must be carefully evaluated to determine which is most appropriate for a specific pipeline construction project.

Tie-in. The most common and the most complex tie-in to be made in offshore pipeline construction is the connection to an abovewater platform on which pumping, compression, oil and gas production, or other equipment is installed. A connection must be made between the pipeline on the sea bed and the equipment on the platform.

Several techniques are used to make this connection (Figs. 7–10 to 7–14). The most common is to place the end of the pipeline on the ocean floor near the leg of the offshore platform to which the riser from the end of the pipeline to the surface of the platform will be attached. A diver then measures the distance accurately from the end of the pipeline to the platform leg. The

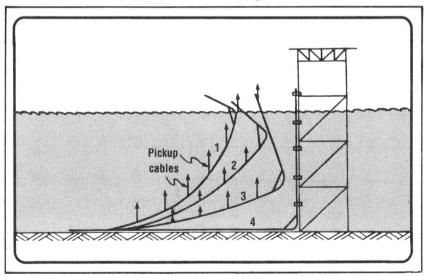

Fig. 7–10.   Conventional riser installation method. Source: *Oil & Gas Journal,* 11 May 1981, p. 105.

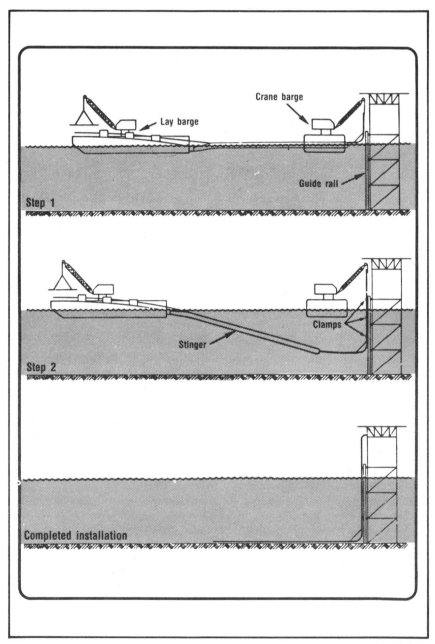

Fig. 7–11.   Guide rail riser installation method. Source: *Oil & Gas Journal,* 11 May 1981, p. 105.

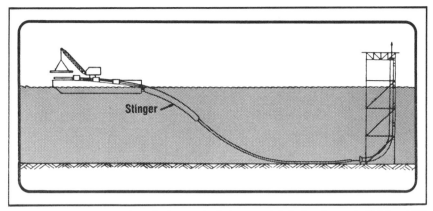

Fig. 7–12.   J-tube riser installation method. Source: *Oil & Gas Journal*, 11 May 1981, p. 105.

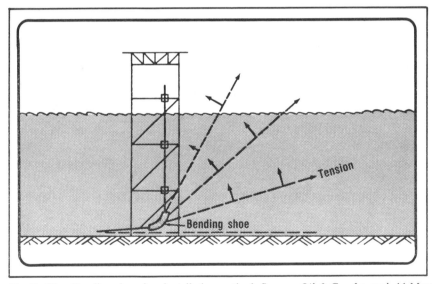

Fig. 7–13.   Bending shoe riser installation method. Source: *Oil & Gas Journal*, 11 May 1981, p. 105.

pipeline is raised to the surface with cables, and pipe is welded on that is of the proper length to reach the platform leg when the pipeline is returned to the ocean floor. A riser bend is then installed to change the pipeline's direction from horizontal to vertical, and sections of pipe are welded onto the bend to provide the vertical connection between ocean floor and platform surface. The required number of sections are welded onto this vertical section as the pipeline is again lowered to the ocean floor.

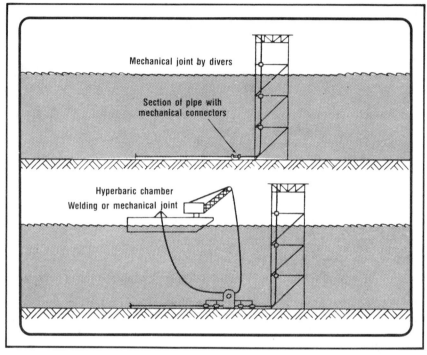

Fig. 7–14.   Underwater riser joining method. Source: *Oil & Gas Journal,* 11 May 1981, p. 105.

This technique for riser installation is limited to water depths of about 300 ft or less with pipelines of 36 in. diameter or more.[15] In deeper water, the length of pipeline that must be raised to attach the riser bend and riser becomes too great for a single barge. The method must also be used in relatively calm sea conditions.

Other riser installation techniques include the guide rail method, the J-tube method, the bending shoe method, and underwater connection by divers.

In the guide rail method, a rail is attached to the platform leg and clamps are installed to allow the riser to slide down the rail. A crane barge or platform-mounted crane picks up the end of the pipeline and is welded to the riser bend. The lay barge begins laying away from the platform, and riser pipe lengths are welded to the riser. After the lay barge has laid enough pipe, the pipeline reaches the ocean floor and the riser is in place from ocean floor to platform surface. The clamps holding the riser to the guide rail are then tightened. This method can only be used when pipeline construction is begun at the platform, and it is also applicable only in relatively calm seas.[15]

In the J-tube method, a J-shaped tube whose inside diameter is larger than the outside diameter of the pipeline to be laid is installed at the bottom of the platform, normally when the platform is being fabricated on land. A cable through the J-shaped section of pipe connects the end of the pipeline on the barge with a pulling device on the platform. When pipeline construction begins, the lay barge remains stationary and the cable is used to pull the end of the pipeline down to the sea floor and up through the J-tube. When the end of the pipeline has reached the platform, the lay barge proceeds laying along the pipeline route.

Because the pipe is bent around the J-shaped path, this method is limited to smaller-diameter lines, usually less than 12 in. It is applicable where pipelaying begins at an offshore platform, but it can be used in more severe sea conditions than other methods. A diver may not be needed during the operation, though one is sometimes used to ensure that the pipeline enters the J-tube properly.

Installing a pipeline riser by the bending shoe method involves installing a curved "shoe" during platform fabrication around which the pipeline will later be pulled. The pipeline is laid past the platform by the lay barge a distance equal to the height of the finished riser. Then the end of the pipeline is bent around the shoe and is raised to vertical by cables while tension is maintained on the pipeline. This method is applicable when pipeline installation terminates at the offshore platform and is considered to be limited to water depths below about 400 ft. [15] Moving the pipe under the bending shoe and aligning it for bending are critical phases of the operation.

Fig. 7–15.   Underwater tie-in assembly. Source: *Oil & Gas Journal,* 13 February 1978, p. 84.

Finally, the pipeline and riser can be installed separately and connected on the ocean floor either by mechanical devices (Fig. 7–15) or by welding. This procedure is more suitable to large-diameter pipelines and can be used when pipeline laying begins or terminates at a platform.

Offshore pipeline burial.   As is the case with onshore pipeline construction, most offshore lines must be buried below the ocean bottom. Burial is necessary to protect the pipeline from damage by ship anchors, fishing gear, and natural hazards. But burying offshore pipelines is more complex than ditching for, and backfilling, an onshore pipeline. First, offshore burying must be done with remotely operated equipment. The sea bed—especially in soft, loose soils—can change quickly, so burial must often be done after the pipeline is in place on the ocean floor.

Government regulations specify how offshore pipelines will be buried in many areas. The most comprehensive regulations (Table 7–2) include those applicable to offshore pipelines in the U.S. Gulf of Mexico, the North Sea, Japan, and Australia. Pipelines installed in the U.S. Gulf of Mexico must be buried to 3 ft below the seabed out to the 200-ft water depth contour, and pipelines crossing an anchorage or fairway area require a 10-ft burial.[16] The U.S. Department of Transportation and the Department of Interior regulate offshore pipelines installed in United States waters; the Bureau of Land Management and the U.S. Geological Survey of the Department of Interior have specified burial requirements in the Gulf of Mexico. In most other offshore areas of the world, burial requirements are usually determined on a case-by-case basis.

North Sea pipeline construction has been subject to various regulations, depending on the country in whose waters the pipeline is located and on site conditions. Early North Sea pipelines were required to be buried with 10 ft of cover, but that requirement exceeded the capability of existing equipment. More flexibility in burial requirements has been the case in recent years.

In Tokyo Bay, Japan, a pipeline was required to be buried to 16 ft, but only 10 ft of cover was obtained. Another pipeline buried to about 8 ft had to be backfilled and the sea bottom restored to its "natural level" to eliminate possible damage to trawling gear.

The two most common approaches to burial are jetting and plowing. Plowing may be done before the pipeline is installed on the ocean floor in certain types of soils, but post-trenching—plowing after the pipeline is in place—is becoming more common as new equipment is developed.

In jetting, a jet sled consisting of a frame on which pumps, jets, and associated equipment are mounted, is lowered to the ocean floor over the pipeline. Its powerful pumps directed under the pipeline force soil from beneath the pipe and allow the pipeline to settle into the resulting ditch. The displaced

TABLE 7–2
Offshore Pipeline Burial Requirements

Country/Agency	Applicable code	Requirements
UNITED STATES		
Department of Transportation (DOT)—Office Pipeline Safety Operations (OPSO)	49 CFR 192 49 CFR 195	Pipeline to be buried below natural bottom.
Department of Interior (DOI) —United States Geological Survey (USGS)	OCS Order 9	No specific requirement.
—Bureau of Land Management (BLM)	43 CFR 2883	Pipeline must be buried to 3 ft below the natural seabed out to a water depth of 200 ft.
UNITED KINGDOM		
Department of Energy (DOE)	Petroleum Pipeline Safety Code 1974 Submarine Pipeline Act, 1975	General guidelines for pipe protection. "The Secretary of State may by regulation make such provisions as he considers appropriate for the purpose of securing the proper construction and preparation in safety operation of pipelines preventing damage to pipelines and securing the safety, health and welfare of persons engaged on pipeline works . . . ."
NORWAY		
Ministry of Petroleum and Energy	Norwegian Petroleum Directorate (NPD), Royal Decrees, 1976	"To the extent reasonable, pipelines shall be protected by burial or by other means to avoid mechanical damage caused by other activities along the route, including fishing and hunting, shipping, and exploration of submarine natural resources. Moreover, the pipelines shall be installed so as not to damage fishing gear."
Industry Recommended Practice	Det norske Veritas (DnV), 1976	"The pipeline is to be supported, anchored or buried in such a way that under the assumed conditions it will not move from its as-installed position, apart from movement corresponding to permissible deformation, thermal expansion, and limited amount of settlement after installation."

TABLE 7–2 (Continued)

Country/Agency	Applicable code	Requirements
NETHERLANDS Inspector General of Mines	Submarine Pipelines for Transport of Gas, 1976	Requirements for burial in shipping lanes or fishing areas to insure safety.
JAPAN Ocean Development Safety Division	Standard for Safety Con- cerning oil and natural gas development, Part 2, Volume 3	General guidelines provided for safety and pipeline stability. However, past experiences have shown that severe burial requirements and possible backfill can be imposed for pipelines crossing areas of fishing activities.
AUSTRALIA Standards Association of Australia	Draft-Australian Standard Rules for Submarine Pipelines, 1974	No specific requirement for burial. Section 5.7—Burying states: "The location of underwater obstructions intersecting the ditch route should be determined in advance of construction activities to prevent damage to such structures. A diver or television inspection shall be made of the ditch ahead of laying operations to insure that the specifications are met."

Source: Mousselli, *Oil & Gas Journal*, 23 June 1980, p. 116

soil then covers the pipe as the jet sled moves along the pipeline. The type of soil on the ocean floor has a great effect on how deep the pipe can be buried in the sea bed by jetting. Where the soil type changes frequently, the pipeline may be left suspended between two relatively hard soil areas. These unsupported spans can cause undesirable stresses in the pipe.

To provide more effective burial under conditions in which jetting is difficult, underwater plows have been developed (Fig. 7–16).[17] A pretrenching plow was first used in the North Sea for lowering a 36-in. loading line between two platforms. The plow was towed by a surface vessel, was about 36 ft long, and weighed 50 tons. In the late 1970s, post-trenching plows were developed. A post-trenching plow is placed on the pipeline on the ocean floor after the pipeline has been laid. In early designs, as the plow was towed, its shares dug into the ocean floor and closed around the pipe, forming a trench under the pipeline. As the plow moved along the pipeline, the pipeline settled into the ditch behind the plow.

A later development was simultaneous plowing in which a plow is positioned on the pipe behind the pipe's touchdown point aft of the lay barge.

Fig. 7–16.   Underwater plow for pipeline burial. Source: *Oil & Gas Journal,* 4 May 1981, p. 133.

Split-share plows, in which share halves are hinged and the force on the leading edge of the plow causes the shares to rotate and close together around the pipe, have also been developed. An example of the split share is one designed for post trenching a 24-in. pipeline in Australia's Bass Strait. The plow is 18 m long and weighs 68 tons. It cut a trench about 1.2 meters deep on that project. Also under development in the early 1980s was a plow that could vary the depth of the trench in uneven sea bottom conditions.

Another approach to offshore pipeline burial has also been used. In one project involving a 36-in. gas pipeline 270 miles long, a program of sandbagging and mechanical backfilling of portions of the line was performed. Backfilling was done by a vessel on the surface that transported excavated material from shore to the pipeline site. The backfill material was fed into a drop pipe that extended to a few feet above the pipeline and delivered the material over the pipeline. Sophisticated navigation equipment was used to ensure that the material was placed over the pipeline. In this project, the average cost of trenching was about $500,000/mile, the cost of sandbagging was about $2.5 million/mile, and the cost of backfilling was an estimated $3 million/mile.[16]

Testing, inspection. Testing of offshore pipelines is similar to that of onshore pipelines. Welds are X-rayed at a station on the lay barge, and the completed pipeline is hydrostatically tested to check for leaks. In addition, because much of the construction of offshore pipelines involves equipment and operations that are remote from view, other methods are also used.

For instance, observation diving bells are used for pipeline route inspection, inspection of the pipeline after it is on the ocean floor, and for other tasks. An example of the use of an underwater vehicle is an observation manipulator bell (OMB) which was used during construction of the trans-Mediterranean pipeline between Tunisia and Sicily (Fig. 7–17).[18] That bell is designed for a crew of two, is equipped with two 5-hp thrusters, and is rated for 3,000-ft water depths. Position and heading are maintained by the pilot's control of the thrusters; vertical position can be controlled either by the bell operator or by the operator on the surface support vehicle. Such a vehicle takes personnel to the ocean floor for direct viewing and work. Operations can be monitored on the surface by real-time video. In the trans-Mediterranean project, the vehicle was used to place and control jacks for the support of the pipeline where it spanned high points in the ocean floor, and to place weight clumps and mats to stabilize the pipeline.[18]

Remote controlled underwater vehicles (RCVs) are playing a greater role in inspection and maintenance of subsea pipelines and offshore platforms. The

Fig. 7–17.   Observation bell can be used for pipeline inspection. Source: *Oil & Gas Journal*, 26 October 1981, p. 153.

RCVs offer the potential for considerable savings over the use of divers. For example, the estimated cost for obtaining samples of sections of an offshore pipeline in Australia's Bass Strait using divers was about $1.2 million, and the job was expected to take several months. The pipeline owner instead invested in an RCV and paid for the vehicle on that one project.

In deep water, the vehicles can be economical. One estimate is that an inspection dive to 800 ft costs about $100,000 for a 10-minute dive, and it may take as much as two weeks to prepare for and perform that dive.[19] An RCV could do the job in 24 hours for about $3,500.

Buckling. An important consideration in designing and installing offshore pipelines, especially in deep water, is the prevention of collapse or buckling. Buckling can be caused by the hydrostatic pressure of water or by longitudinal bending of the pipe during installation. The danger of collapse or buckling increases with greater water depths and larger pipe. Factors affecting the tendency for buckling and collapse include pipe out-of-roundness, ratio of diameter to wall thickness, yield strength, and stress/strain behavior of the pipe steel.[20] In conventional pipelaying methods, the pressure inside the pipe during installation is atmospheric, and the external hydrostatic pressure can cause severe stresses. In some pipelay methods, pipe has been pressurized internally to offset the hydrostatic pressure.

The seriousness of pipe buckling depends on the net external pressure on the pipe relative to the buckle initiation pressure and the buckle propagation pressure. The initiation pressure is higher than the propagation pressure.

Collapse of a submarine pipeline may be local only or it may "propagate" itself along the pipeline, damaging a significant length. The greatest danger of collapse is during installation, when the pipe is subjected to laying stresses. But it can also occur after the pipe is installed if, for example, there is no internal pressure on the line.

To avoid collapse and buckling requires careful design, control of curvature of the pipe during installation to stay within design criteria, and quality control of the pipe to avoid lengths with excessive out-of-roundness or other imperfections.

Pipe buckling can be expensive to repair. A dry buckle may take several days to repair; a wet buckle (where the pipe is flooded with water) can take weeks to repair. Repairs must normally be done with the pipelay barge on location, and lay barge rental can run as high as several hundred thousand dollars per day.

In addition to design and quality control, equipment is available to use during construction to detect buckles and to halt the propagation of a buckle should one occur. Even though a properly designed, carefully installed pipeline is not likely to buckle, the severe consequences of a buckle make it common to install buckle arrestors when installing pipelines in deep water.

There are three types of buckle arrestors: the free-ring arrestor, heavy-walled cylinder or integral arrestor, and the welded ring arrestor.[20] The free-ring arrestor is a steel sleeve of larger-diameter pipe that is slipped over the pipe joint. The integral arrestor is of heavier wall thickness than the pipe but usually has the same inside diameter and is welded into the pipeline. The welded ring arrestor is similar to the free-ring arrestor but is welded to the pipe.

Anchoring. Offshore pipelines are often anchored to the seabed to prevent movement due to current and other forces. One mechanical anchoring method involves screwing auger-like anchors into the sea bed that hold a bracket down over the pipe, pinning the pipe to the ocean floor. Auger-type anchors must be used in a soil that offers sufficient resistance to the anchor in order to be effective. Other types of anchors that depend on weight—gravity anchors—have also been used. They can be either set on the pipeline or bolted to the pipeline and are typically made of concrete.

Various techniques have developed for installing pipeline anchors. The type of anchor and installation method must be designed for each specific location. Soil conditions, particularly soil resistance, and the seabed profile are key factors in choosing the proper type of anchor, the placement method, and the location where anchors will be needed.

Arctic pipeline construction

Arctic pipeline systems, both onshore and offshore, require special approaches to design, installation, and operation. Logistics is a key problem, and offshore, much of the Arctic area is covered by ice most of the year. Arctic pipeline designs therefore often emphasize prefabrication as a way to reduce cost and meet construction schedules. Work productivity is also a consideration in all Arctic projects. It must be kept in mind that many operations will have to be performed by a worker dressed in heavy clothing.

Pipelines designed for these severe environment areas differ from conventional pipeline systems in more moderate land and offshore areas:

1. Special pipe steel may be required to withstand the low temperature.
2. On land, pipelines must be designed to prevent damage to the pipeline and to the environment in permafrost areas. Where there is no permafrost, frost heave must be considered.
3. Offshore, special pipe laying techniques must often be used.
4. Offshore pipelines must be protected from damage by moving ice masses that contact the sea bed.
5. A narrow operating temperature range may require special equipment and operating procedures.

6. Startup of crude oil pipelines after a shutdown can be a critical operation; provisions must be made in the design for this occurrence.

Onshore construction. Construction of pipelines in Arctic onshore areas is dictated by a design aimed at preventing damage to the pipeline and the environment due to the melting of the permafrost and preventing damage to the pipeline due to frost heave. Permafrost is a layer of the earth's surface that remains below 32°F for extended periods, sometimes defined as two years or more, regardless of the season. If allowed to thaw, its volume decreases and movement of the pipeline can cause pipe stress to exceed design limits. About half the length of the 48-in. diameter trans-Alaska crude pipeline was installed aboveground on specially designed pipe supports through permafrost areas to solve this problem. Melting of permafrost areas can also cause unwanted effects on the environment.

Natural gas pipelines laid through permafrost areas may have to be designed so the gas can be cooled to maintain a temperature low enough to prevent permafrost melting. But this leads to another problem in areas where the soil is not frozen: frost heave. The cold gas can freeze the soil, causing an increase in volume that again can put stress on the pipe above design limits. This heave has been found to be greater than that which would occur due solely to the freezing of the water in the soil; additional water is attracted to the area of freezing, aggravating the heave conditions.

Insulation can be used to limit the size of the frost-heave area, or the trench can be overexcavated and the original material replaced with a granular material that is not frost susceptible. Another approach is to use a computer model to predict heave and the resulting stresses on the pipe. At locations where heave does not cause undue stress, no measures are taken. Where design stresses are exceeded, preventive measures can be included in the design. The proposed Alaska Natural Gas Transportation System was designed to operate at a maximum temperature of 28°F and a minimum of 0°F. The maximum was set to prevent thawing of the permafrost, and the minimum limit was required to remain within fracture control limits of the pipe steel and avoid condensation of liquids in the pipeline.[21]

In all Arctic construction, logistics and working conditions dictate many construction procedures. For the trans-Alaska crude pipeline, for example, about 3 million tons of materials had to be delivered to the project and construction camps had to be built for 23,000 personnel.

Offshore construction. Many construction techniques in Arctic offshore areas are similar to those used in more moderate locations. But there are also differences required to cope with ice and with short open-water periods. For instance, a conventional lay barge may be costly due to the short work periods.

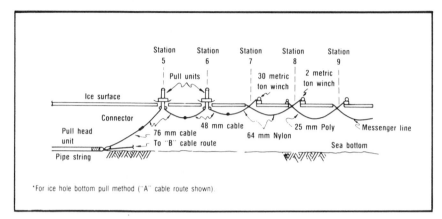

Fig. 7–18. Method for laying pipeline under ice. Source: *Oil & Gas Journal*, 21 September 1981, p. 145.

Bottom tow methods have advantages because much of the work can be done in winter and the lines placed during the open-water season. Laying pipe with a reel barge could also offer advantages in Arctic waters.

On-ice construction methods are also possible. A pipeline may be laid in very shallow waters by cutting a trench through the ice and into the soil, much as in conventional land pipeline construction. In deeper water, the pipe can be welded together on the ice and lowered to the sea bed through a slot in the ice.[22] Modifications of the bottom-pull method can also be used in ice-covered areas. In the ice-hole bottom-pull method (Fig. 7–18), long pipe strings are pulled into place by cable-pulling units from a series of holes in the ice.[23]

Pipe in Arctic waters must also be protected from damage by moving ice masses that may gouge the ocean floor. If the depth that these masses penetrate the seafloor is known, one approach is to trench the pipe below that depth. The trenching method used—jetting, dredging, plowing—depends on the type of soil encountered. Those trenching methods with higher rates are the most desirable because of the short open-water season, provided the method used is able to trench in the type of soil that exists along the pipeline route.

REFERENCES

1. Earl Seaton, "U.S. Pipelines Keep Energy Moving," *Oil & Gas Journal*, (22 November 1982), p. 73.

2. G. Alan Petzet, "Northern Tier to Persist Despite New Blow," *Oil & Gas Journal*, (7 September 1981), p. 54.

3. Howard Wilson, "Sohio Gives up on West Coast-Texas Line," *Oil & Gas Journal*, (19 March 1979), p. 66.

4. John P. O'Donnell, "Coal-Tar Enamel Remains Most-Preferred Pipe Coating," *Oil & Gas Journal*, (6 July 1981), p. 120.
5. G. Alan Petzet, "Pipeline Record Claimed in Orinoco Crossing," *Oil & Gas Journal*, (19 April 1982), p. 54.
6. William R. Kozy and Keith J. Meyer, "Pipeline Suspension Bridge Required Precise Design," *Oil & Gas Journal*, (30 June 1980), p. 63.
7. John M. Odusch, "Anchoring Pipelines Prevents Damage," *Oil & Gas Journal*, (24 November 1980), p. 101.
8. P.R. Scott and Kor N. Zijlstra, "Flags Gas Line Sediment Removed Using Gel-Plug Technology," *Oil & Gas Journal*, (26 October 1981), p. 97.
9. "Why and How to Dry Gas Pipelines," *Pipe Line Industry*. (October 1981), p. 35.
10. Bill Smith, "Offshore Pipeline Construction—1: Offshore Line Construction Methods Examined," *Oil & Gas Journal*, (4 May 1981), p. 154.
11. "Record Trans-Med Pipelines Finished," *Oil & Gas Journal*, (9 February 1981), p. 39.
12. Svend Jorgensen, "Flow Lines Laid by Reel-Ship *Apache*," *Oil & Gas Journal*, (5 May 1980), p. 160.
13. "Unique Pipe-Laying Vessel's First Job in North Sea," *Oil & Gas Journal*, (28 June 1976), p. 72.
14. N.L. Fernandez, "Tow Techniques for Offshore Pipelaying Examined for Advantages, Limitations," *Oil & Gas Journal*, (22 June 1981), p. 64.
15. Bill Smith, "Offshore Pipeline Construction (Conclusion): A Look at Pipeline Riser Installation Techniques," *Oil & Gas Journal*, (11 May 1981), p. 105.
16. Al H. Mousselli, "Government Regulations for Offshore Pipeline Burial Vary by Design," *Oil & Gas Journal*, (23 June 1980), p. 116.
17. R.J. Brown, "Use of Underwater Plowing May Cut Offshore Pipelaying Costs," *Oil & Gas Journal*, (4 May 1981), p. 133.
18. Robert E. Lewis, "Observation Manipulator Bell Proves Worth In Trans-Mediterranean Pipeline Construction," *Oil & Gas Journal*, (26 October 1981), p. 153.
19. "Offshore Construction Industry Cutting Costs with RCVs," *Oil & Gas Journal*, (23 November 1981), p. 48.
20. B.K. Jinsi, "Collapse and Buckling Strength Considerations Are Pinpointed for Offshore Pipeline Design," *Oil & Gas Journal*, (3 May 1982), p. 217.
21. Nils Hetland, "Design Considerations Included Environmental Protection for Alaska Segment of ANGTS," *Oil & Gas Journal*, (18 October 1982), p. 124.
22. W.J. Timmermans, "Design of Offshore Pipelines for Ice Environments," ASCE Annual Convention and Expo, 25–29 October, 1982, New Orleans.
23. O.M. Kaustinen, "Pipelining Gas From the Canadian High Arctic," *Oil & Gas Journal*, (21 September 1981), p. 145.

8

WELDING TECHNIQUES AND EQUIPMENT

THE overwhelming bulk of oil and gas pipeline construction is done by
welding the individual joints of pipe together. However, other types of
connections are used, including threaded couplings and mechanical connectors.

Very strict controls on pipeline welding require that both a welding
procedure and the welders who will use the procedure be qualified by testing.
Comprehensive inspection of completed welds is also required, and the causes
of weld defects and their prevention continues to be the focus of much study.

The aim of detailed specifications and regulations, and further research and
development work on welding procedures and equipment, is to ensure that oil
and gas pipelines are safe.

It is not possible to detail all aspects of pipeline welding in this short chapter.
However, a brief overview of welding processes, procedures, and equipment,
and highlights of applicable regulations, will indicate the complexity and
sophistication of modern pipeline welding.

This chapter discusses only field pipeline welding in which a circumferential
weld is made to join individual lengths of pipe. Welding performed during pipe
manufacture—longitudinal and spiral welds—was discussed in Chapter 3.

Welding processes

In a broad sense welding is "a metal-joining process wherein coalescence is
produced by heating to suitable temperatures with or without the application of
pressure and with or without the use of filler metal."[1] The sources of heat for
welding include electric arc, electric resistance, flame, laser, and electron beam.

The first three are traditional methods; laser and electron beam welding are relatively recent developments.

Most processes used in field pipeline welding use a filler metal, do not involve the application of pressure and depend on an electric arc for the heat source.

Shielded metal arc welding. The heat for this process is provided by an electric arc that melts a consumable electrode and some of the metal being welded. When the weld metal cools, it hardens to form the weld. The consumable electrode is melted continuously by the heat of the electric arc. As in all arc welding processes, the electrode serves as one pole of the arc; the steel being welded is the other pole. The electrode, the steel pipe, and the arc make up an electric circuit: the welding circuit.

A covered electrode has a solid metal core and an outer layer of material that insulates the core from accidental contact with the pipe. The core covering also provides the gas to shield the weld from air and may contain special elements to improve weld quality.

Submerged arc welding. In this process, too, heat is supplied by an electric arc and a consumable electrode is used. In this technique, however, a granular flux composed of silicates and other elements is deposited on the weld joint. The arc melts some of the flux and is submerged in the liquid slag that is produced by this melting. The electrode in this method is wire that is fed continuously to the weld joint. High currents used in this technique allow the weld to penetrate deeper below the surface of the pipe than is possible with other welding processes.

Gas-metal arc welding. This process also uses the heat from an electric arc. The arc is covered by an inert gas, such as argon or helium. The inert-gas-shielded metal arc process uses a consumable, continuous electrode. Since this process requires no flux, no slag is produced on top of the weld. Gas for shielding is delivered to the weld area through a tube; the electrode is fed down through a guide within the tube. Gas-metal arc welding (GMAW) is particularly applicable to welding difficult metals and alloys that are susceptible to contamination from the atmosphere, and porosity.

Carbon dioxide welding is similar to gas-metal arc welding except that CO_2 is used as a shielding gas.

Gas-tungsten arc welding. An inert gas shield is required when welding with tungsten electrodes using the gas-tungsten arc welding (GTAW) process. This process is particularly suited to welding thin material and to depositing the first weld bead (root pass) because penetration can be controlled more easily than with other welding processes. Good heat control is possible with this process, and it is possible to weld with or without filler metal. The nonconsumable electrodes are not deposited as part of the weld metal. The steel being welded is melted, and the electrode serves only as one pole of the electrical circuit. In some processes, however, a filler wire can be fed into the weld joint if additional metal is needed to fill the joint.

There are also other arc welding processes:

1. *Plasma arc welding,* in which a plasma is produced by the heat of a constricted arc/gas mixture. This type of welding is similar to tungsten arc welding since an inert gas is used, but plasma arc welding's constricting orifice is unique. Plasma/metal-inert-gas (MIG) welding combines the features of plasma arc and inert-gas metal arc processes and permits deep penetration when welding thick material or higher speeds when welding thinner material.
2. *Flux-cored arc welding,* which is similar to submerged arc and shielded metal arc welding, except the flux is contained in a metal sheath instead of on the wire.
3. *Electroslag welding,* which is begun much as conventional submerged arc welding. When a layer of hot molten slag is formed, arc action stops and current passes from the electrode to the work through the slag. Heat generated by the resistance to current through the slag fuses the edges of the work pieces.

Electron beam welding. Though not common in pipeline welding in the early 1980s, the electron beam process could have application in proposed laying techniques for installing offshore pipelines in very deep water. The J-curve pipelay method, discussed in Chapter 7, involves lowering pipe from near vertical rather than in a horizontal S-shaped configuration used in conventional offshore pipelaying.

In the electron beam welding process, coalescence is obtained by concentrating a beam composed primarily of high-velocity electrons impinging on the surfaces to be joined. Electrons accelerated by an electric field to extremely high speeds and focused to a sharp beam by electrostatic or electromagnetic fields provide heat for welding.

Because the J-curve pipelay technique requires welding to be done at a single station rather than the several stations common on a conventional lay barge, electron beam welding offers an advantage. It is quicker than other methods and can weld thick-wall pipe in a single pass. The process also requires no preheating or postheating of the weld area.[2] In tests on 24-in. pipe with a wall thickness of 1.2 in., complete welds were made in less than 3 minutes. Conventional welding of a joint in the same pipe would require 1½ hours.

Accurate positioning of the beam is important in the electron beam process, and a vacuum must be maintained in the chamber around the pipe. The beam must be concentrated on a spot about 1 mm in diameter, and tolerances are critical.

Welding procedures and equipment

Most welding processes can be used with either of the two general types of welding procedures: manual or automated. Of all of the oil and gas pipeline welding done, the bulk is done by manual welding procedures. In manual welding, the welder holds the electrode. He must be highly skilled at maintaining the proper distance between the electrode and the pipe and in moving the electrode along the weld seam at the proper rate. Automated welding equipment also requires skilled operators, but it is easier to obtain consistent, uniform welds because the welding electrode is moved mechanically at the desired distance from the seam and at the optimum speed.

Weld passes. Whether done by manual or automated welding, each pipeline circumferential, or *girth,* weld is completed by making several passes around the circumference of the pipe at the seam between the two joints. The number of passes, or beads, depends on the thickness of the pipe wall and the welding procedure. Regardless of pipe wall thickness, however, a pipe weld consists of a root pass, the first pass made around the seam; a hot pass applied over the root pass; several fill passes, which fill in the space between the pipe ends; and a cap pass.

The root pass is especially critical. An improperly applied root pass can burn through the pipe wall, producing a defective weld that will cause the weld to be rejected when inspected. Since the root pass initially joins the two lengths of pipe, it is necessary that they be correctly aligned and spaced before welding begins. In some procedures, the root pass is deposited from inside the pipe; then all other weld passes are made from the outside of the pipe. In other cases, the root pass is made from the outside of the pipe. If a metal slag is produced during welding, it must be removed before the next weld pass is made.

Following the root pass, the hot pass is deposited. It is normally made before the root pass bead cools completely. The hot pass is also critical to overall weld quality. After it is made, the hot pass is cleaned of slag, if necessary, before fill pass welding begins.

After the hot pass, the required number of fill passes is made. Fill passes build up the weld and fill in the seam between the two pipe joints with weld metal. The number of fill passes required depends primarily on the wall thickness of the pipe being welded. It also may depend on the particular welding procedure being used. Thin-wall pipe may only require one fill pass, while the

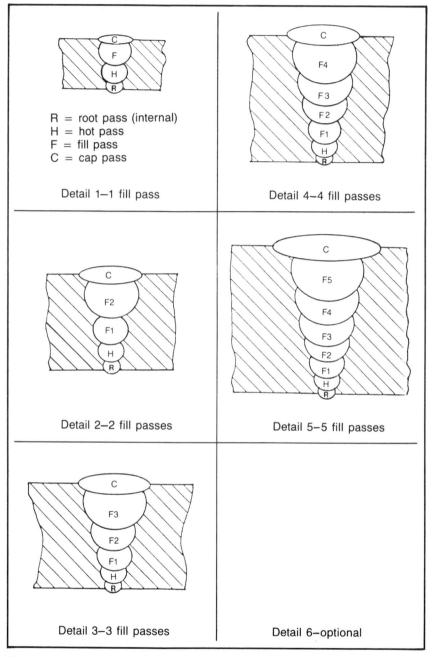

Fig. 8–1.   Weld passes. (Courtesy Crutcher Resources).

number of passes may be five or more for pipe with a thicker wall. The speed with which pipeline welds can be made depends heavily on the number of fill passes required, in addition to the outside diameter of the pipe. A welder can deposit a certain amount of weld material in a specified time. If a number of passes are made on the same weld seam, fewer complete welds can be made per day. The amount of weld material that must be deposited also increases as the pipe diameter increases, so fewer welds can be made in a day on large-diameter pipe.

The final weld is the cap pass. It is normally wider than the last fill pass, and weld metal is not deposited in as thick a layer as is the case with the other weld passes. The top of the cap pass extends slightly above the pipe exterior.

Fig. 8–1 shows examples of the weld passes required in a pipeline girth weld.

Manual welding. The largest share of oil and gas pipeline welding is still done manually (Fig. 8–2). Welders along the pipeline right of way use vehicle-mounted welding machines and weld the individual joints of pipe together alongside the pipeline ditch. The number of welders on a pipeline job depends on the length of the pipeline, the diameter and wall thickness of the pipe, and other factors.

Fig. 8–2. Manual pipeline welding.

Typically, one welder applies the root pass, another welder the hot pass, one or more welders the fill passes, and another welder the cap pass.

Manual pipeline welding not only requires great skill, but conditions are often difficult and unpleasant. Since the pipe seam must be welded completely around the circumference of the pipe with the pipe in a horizontal position, the welder must be down on the ground in an uncomfortable position to complete the weld on the lower portion of the pipe. In extreme weather, it is often necessary to provide a shelter or enclosure for the welder to protect against wind, blowing dirt or sand, and cold. Not only is this needed for the welder's comfort, but blowing dirt, moisture, and high wind can reduce weld quality. Some welding shelters include wooden floors, doors, exhaust vents, lighting, and heating to provide properly controlled welding conditions. Other shelters consist only of a canvas hood over the welder and the joint being welded.

Automatic welding. In the late 1960s, automated welding equipment reached the commercial development stage (Fig. 8–3). Since that time, it has

Fig. 8–3. Automated welding on offshore pipeline. Source: *Oil & Gas Journal,* 10 January 1977, p. 91.

been used in pipeline welding in many parts of the world for a wide range of pipe sizes. It is used for both land pipeline construction and on offshore lay barges for offshore pipeline construction. One firm that offers automated pipeline welding equipment and expertise had used its equipment to weld over 9,000 miles of pipeline by 1982, ranging in size from 16 in. diameter to 60 in. diameter. Wall thickness ranged up to more than 1 in.[3]

These advantages of automated welding systems have been cited:

1. Increased weld metal deposition rate
2. Reduced volume of weld metal
3. Improved consistency of weld strength, toughness, and radiographic quality
4. Reduced vulnerability of weld quality to human error
5. Reduced physical strain on the welder/operator
6. Ease of training operators
7. Reduced manpower and equipment requirements for heavy wall and large-diameter pipe

In automated welding, as in manual welding, a root pass, hot pass, fill passes, and cap pass are required. A continuous wire is used to supply the weld metal to be deposited.

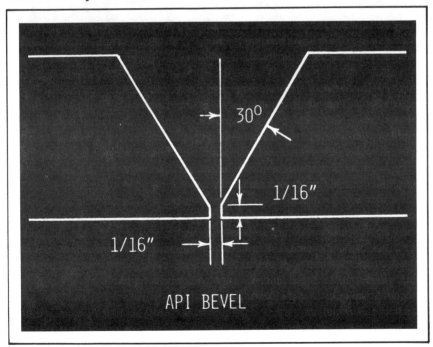

Fig. 8–4.   Standard API bevel. (Courtesy Crutcher Resources).

The CRC automated welding system, for example, is a fine-wire, gas-metal arc welding system consisting of three major components: a pipe end facing machine, a combination internal welder/line-up clamp, and an external welding carriage.

The pipe facing machine serves a special purpose. Line pipe is usually manufactured and delivered with the ends bevelled at a standard 30 degrees (Fig. 8–4). For automated welding, a modified bevel has been found to increase the quality of the weld, and the facing machine is used to modify the standard 30-degree bevel in the field to prepare for automated welding. (Details of the modified bevel will be discussed later in this chapter.) The facing machine (Fig. 8–5) has a clamp section and a machining section. The clamp section secures the machine to the end of the pipe joint and aligns the machine with the end of the pipe. Cutting tools mounted on a rotating face plate then machine the end of the pipe to the desired bevel. The pipe facing machine typically is suspended from a sideboom tractor, and the facing operation normally takes from two to five minutes.[3]

Fig. 8–5.   Pipe facing machine. (Courtesy Crutcher Resources).

The internal line-up clamp/welder (Fig. 8–6) positioned inside the pipe aligns the two pipe ends, locks them in place, then automatically welds the root pass on the inside of the pipe. The welding section of the unit has four welding heads for pipe diameters from 24 in. to 38 in. and six heads for 40-in. to 60-in. diameter pipe. The heads are mounted around a ring gear driven by an electric motor. A four-head machine, for example, begins welding with two heads at the

Fig. 8–6. Internal clamp/welder. (Courtesy Crutcher Resources).

12 and 3 o'clock positions as seen from the open end of the pipe. These heads weld downhill to 3 and 6 o'clock, respectively, typically at about 30 in./min. The other two heads move into position at 12 and 9 o'clock, and when the first two heads are finished, the second two weld from 12 to 9 o'clock and from 9 to 6 o'clock, respectively. When the root pass bead is complete, the clamps are removed and the unit propels itself through the pipe joint just welded and stops automatically at the open end.

The external welder carriages (Fig. 8–7), called *bugs*, make the remaining welds—hot pass, fill passes, and cap pass—from the outside of the pipe. For the different passes, the bugs have different gas shielding nozzles, travel speeds, and welding tip oscillation.

The bugs travel on spring steel bands attached to the pipe after the new bevel is cut and before the pipe joint is welded into the pipeline. Each bug has a carriage section, control box section, and welding section. The carriage, which can be adjusted for pipe diameter, is attached to the steel bands; the control box carries electronics that control travel speed, wire feed speed, welding tip oscillation frequency, and wire and gas shutoff delay. The welding section consists of the welding tip, wire feed drive motor, oscillation motor, gas shielding nozzle, and welding wire spool.

Bugs are used in pairs, each making half a weld pass from 12 o'clock to 6 o'clock, one in a clockwise direction and one in a counterclockwise direction. Hot pass bugs begin welding before the internal root pass is completed and typically travel at 40–45 in./min.

Fill pass bugs begin at the same time but not at the same point. Starting positions are changed and reversed on alternate fill passes to avoid overlapping starts and stops.

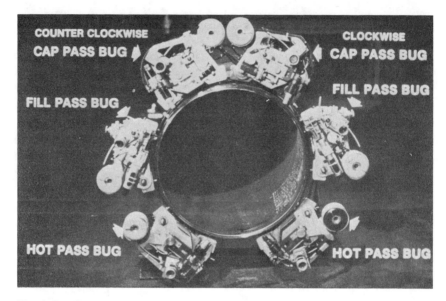

Fig. 8–7. External welding carriages. (Courtesy Crutcher Resources).

Modified bevel.   CRC uses an end bevel that differs from the standard end preparation and line-up in two respects (Fig. 8–8): the angle of the bevel and its shape are different and the two pipe ends are butted together rather than being positioned with a space between them.

The standard 30-degree bevel was changed for several reasons. If the pipe is not perfectly round when the bevel is cut, the cutting tool produces variations in the root face thickness. Also, the internal line-up clamp may round out the pipe ends, distorting the originally flat plane of the bevel. These conditions can produce problems in automatic welding. In addition, the standard bevel requires a larger volume of weld metal to fill the seam. Even for manual welding, a modified bevel is often used on heavy-wall pipe to reduce the amount of weld metal required.[4] The elimination of the gap between the two pipe ends in the modified bevel configuration not only reduces the volume of weld metal required but can reduce alignment time and burn-throughs.

Onshore/offshore

The same welding processes can be used for both land and offshore pipeline construction. Both manual and automated welding procedures are used for land and offshore construction.

On land, welders move along the pipeline right of way as the pipeline is constructed. The root pass welder moves from one weld seam to the next,

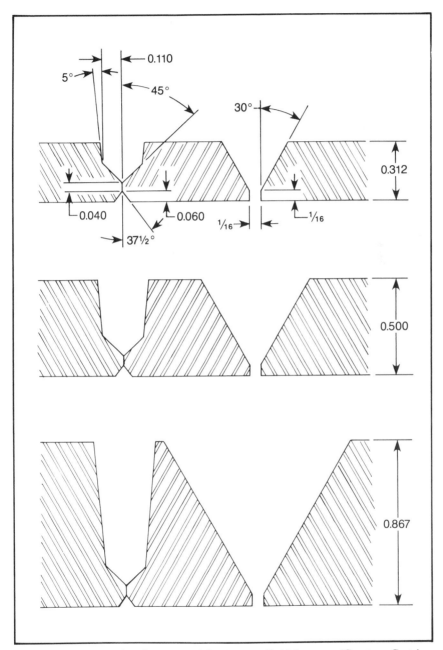

Fig. 8–8. CRC bevel and API bevel for three wall thicknesses. (Courtesy Crutcher Resources.)

followed by the hot pass welder, fill pass welders, and the cap pass welder. In offshore construction, the welding stations are stationary (Fig. 8–9) on the lay barge, and as the lay barge moves along the pipeline route, the successive pipe joints move through each welding station. The number of weld stations on a lay barge varies, and it also contains weld-inspection stations and coating stations.

Fig. 8–9. Welding on offshore pipelay barge. Source: *Oil & Gas Journal,* 30 April 1979, p. 158.

Double jointing. Often a pipeline construction project can be completed faster if only every other weld seam must be completed on the pipeline right of way. This is often the case where right-of-way conditions are difficult or where weather or other environmental conditions are severe. By welding two joints of pipe together in a fabrication yard or building (Fig. 8–10) and transporting the double-length joints to the right of way, field construction time can be reduced. At the double-jointing yard, welding stations are stationary, and necessary shelters or other facilities can easily be built. Welding is faster than on the right of way, where difficult environmental conditions exist and weld quality can often be improved.

Fig. 8–10. Double jointing underway at pipe yard. Source: *Oil & Gas Journal*, 15 October 1979, p. 117.

Double-jointing is done on both land and offshore pipeline construction projects.

Regulations

Part 192 of the U.S. Code of Federal Regulations, Title 49 (49 CFR)—Transportation—sets out welding regulations for natural gas pipelines in subpart E. Covered are the qualification of welding procedures, qualification of welders, preparation for welding, preheating, stress relieving, inspection and testing of welds, and other aspects of the circumferential weld between two joints of line pipe. These regulations do not apply to welding done during pipe manufacture.

Procedure qualification. Welding of pipelines in the United States must be accomplished using written welding procedures as outlined in 49 CFR. These procedures are qualified under the American Society of Mechanical Engineers (ASME) Boiler and Pressure Vessel Code or API standards. Certain steels— carbon steels with a carbon content of less than 0.32% or with a carbon equivalent (carbon plus one-fourth manganese) of less than 0.65%, and certain alloys whose weldability is similar to these carbon steels—require separate qualification of welding procedure. Each welding procedure must be recorded in detail during the qualifying tests and the record retained whenever the procedure is used.

Welder qualification. In 49 CFR, each welder is required to be qualified according to the Boiler and Pressure Vessel Code or API standards. Steels are divided into the same general group as is the case for procedure qualification as far as the need for separate qualification is concerned.

Other welder qualification rules depend on the operating conditions of the pipeline to be welded, the time since the welder last was qualified on the procedure, the results of tests on a weld made by the welder, and the pipe size on which the welder is qualified. Rules also specify limitations on welders, including who may weld compressor station pipe and components.

Different welding qualification tests are required for different purposes. The basic test for welders qualifying to weld low stress level pipe, however, serves as an example of key points in qualification tests.[5] Under the rules, that test must be made on a pipe 12 in. or less in diameter. The pipe must be welded in a horizontal fixed position, so the test includes at least one section of overhead welding. Bevelling, root opening, and other details of the test must conform to the procedure under which the welder is being qualified.

When the test is completed, the test weld is cut into four coupons and is subjected to a root bend test. If two or more of the coupons develop a crack that is more than 1/8 in. long in any direction in the weld material or between the weld material and the base metal, the weld is unacceptable. Other qualification tests contain other requirements, but the general approach to testing is similar.

Preparation for welding. Safety regulations require that the welding operation be protected from weather conditions that would impair the quality of the completed weld. Before beginning, welding surfaces must be clean and free of any material that may be detrimental to the weld. Also, pipe sections must be properly aligned for depositing the root bead.

Specifications require that carbon steel with a carbon content in excess of 0.32% be preheated for welding; in some cases, steels with lower carbon contents must also be preheated. When steels with different preheat temperatures are being welded, the higher of the two preheat temperatures must be used. Preheat temperature must be monitored to ensure the proper temperature is reached and maintained during welding.

Stress relieving. Most welds on steels with a carbon content of more than 0.32% or a carbon plus one-fourth manganese content greater than 0.65% must be stress-relieved according to procedures in the ASME Boiler and Pressure Vessel Code. Stress relieving may also be required on steels with less carbon content if conditions exist that cool the weld too fast for good weld quality.

Stress relieving, as the name suggests, is a treatment used to relieve internal stresses in the vicinity of the weld that result from the heat of welding. Because the weld metal cools quickly if temperature is not controlled, these stresses occur. Stress relieving involves heating the weld area after the weld is complete and letting it cool slowly.

Stress relieving requirements also depend on pipe wall thickness; the type of connection, such as branches, slip-on connections, flanges, etc.; welding of dissimilar materials; and other factors. Stress relieving done under 49 CFR must be accomplished at a temperature of at least 1,100°F for carbon steels and at least 1,200°F for ferritic alloy steels. When stress relieving a weld between materials with different stress-relieving temperatures, the higher temperature must be used. Temperatures must be monitored during stress relieving to ensure that they are uniform and that the proper stress relieving cycle is performed.

Inspection, testing. Pipeline welds must be inspected visually to ensure they are performed in accordance with the welding procedure and that the welds are acceptable under the appropriate specifications. In the United States, to comply with provisions of 49 CFR, nondestructive testing is also required on welds made on a pipeline that will be operated at a pressure that results in a hoop stress of 20% or more of the specified minimum yield strength (SMYS). There are two exceptions to this nondestructive testing requirement. Nondestructive testing is not required even though hoop stress will exceed 20% of SMYS if 1) the pipe has a nominal diameter of less than 6 in. or 2) the hoop stress will be less than 40% of the SMYS and the number of welds is so small as to make such testing impractical. Nondestructive testing must be performed according to written procedures and by specially trained personnel.

Radiography (X-ray) is the most widely used method of testing pipeline welds nondestructively. In radiography, a picture is made on a sensitized film by exposing the film to X-rays (Fig. 8–11). Defects in the weld, such as cracks, porosity, or slag inclusions, appear as spots or lines on the developed film.

The number of welds in a pipeline that must be X-rayed depends on the type of line and the area in which it is located. Detailed descriptions of different types of locations are included in CFR 49. In general, Class 1 locations are those that contain few buildings. Classes 2, 3, and 4 become progressively more densely populated. Class 4 contains buildings with four or more stories. The class designation is made based on an area that extends 220 yards on either side of the centerline of any continuous 1-mile length of pipeline.

In these classes, nondestructive testing of natural gas pipeline welds is required as follows:

1. In Class 1 locations, except offshore, at least 10% of the welds must be X-rayed.
2. In Class 2 locations, at least 15% of the welds must be X-rayed.
3. In Class 3 and 4 locations, at crossings of major or navigable rivers, and offshore, the regulations in 49 CFR state that 100% of the welds must be X-rayed if practicable but not less than 90%.
4. Within railroad or public highway rights of way, including tunnels, bridges, and overhead road crossings—and at pipeline tie-ins—100% of the welds must be inspected.

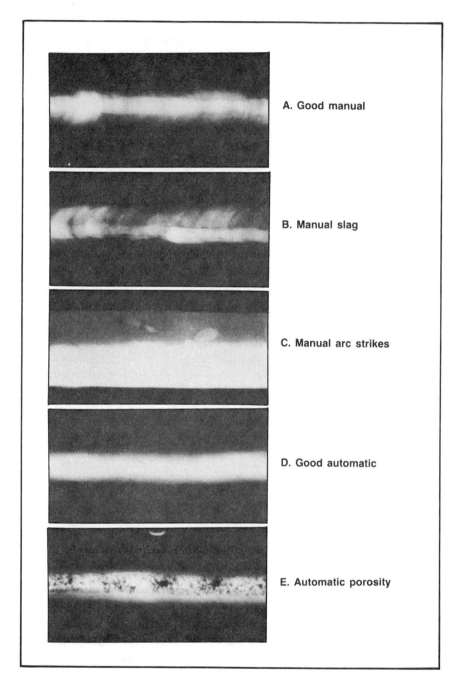

Fig. 8–11.   Radiographs of welds. Source: *Oil & Gas Journal*, 30 March 1981, p. 155.

Other requirements also are specified in 49 CFR for nondestructive testing. For instance, "except for a welder whose work is isolated from the principal welding activity, a sample of each welder's work for each day must be nondestructively tested. . . ."

Records of nondestructive testing must be retained that show by milepost, engineering station, or geographic feature the number of girth welds made, the number nondestructively tested, the number rejected, and the disposition of rejects.

Welds that are unacceptable must be removed or repaired. Under 49 CFR specifications for natural gas pipelines, for example, a weld must be removed if it has a crack that is more than 2 in. long or that penetrates either the root or second bead. Each weld that is repaired must have the defect removed down to clean metal, and the segment to be repaired must be preheated. After repair, the weld must again be inspected to ensure it is acceptable.

Regulations in 49 CFR Part 195 for liquids pipelines specify similar procedures for pipeline welding. Welds must be performed according to written procedures, and welders must be qualified. Nondestructive testing of welds in liquids pipelines is also required. At least 10% of the welds made by each welder during each welding day in a liquids pipeline must be nondestructively tested over the entire circumference of the weld. In addition, many locations require that 100% of the welds be nondestructively tested. Those instances include any onshore location where loss of liquid from the line could pollute a body of water and any offshore area, within railroad or public road rights of way, at overhead road crossings and in tunnels, at pipeline tie-ins, within any incorporated subdivision of a state government, within populated areas, and when installing used pipe.

Regulations being reviewed. In the early 1980s, portions of the Materials Transportation Board's safety standards for welding high-pressure liquids and natural gas pipelines in the United States were being reviewed. The goal was to update, simplify, and make less costly the pipeline safety standards.[6] Areas under study at that time included the following:

1. Repair and removal of girth weld defects
2. Retention of radiographic film of girth welds in liquids pipelines
3. Use of fracture mechanics criteria as an alternate weld acceptance standard
4. Other regulations involving nondestructive testing, qualification of welders, stress relieving, and arc burns

Weld defects.[7] The expense of cutting out unacceptable welds makes it necessary to use techniques that will minimize pipeline weld defects. Proper joint preparation is the first step in preventing weld defects. The ends of the pipe must be clean and, if bevelled, must have proper angle and thickness of bevel.

The gap between the ends of the pipe, if specified, must be that prescribed for the pipe size and welding method used, and the two joints must be properly aligned before welding begins. Other factors affecting weld quality include proper welding current and proper electrode angle.

Internal undercut, a common defect, can be due to poor fit-up or poor joint preparation. This defect can occur if the root face is too small, the root opening is too large, or the welding current is excessive. It can also result when an internal chamfer has been made on the end of the pipe when improperly removing burrs from the inside edge of the pipe end.

Sidewall undercut, or wagon tracks, can occur when the first weld pass is made. If the root pass bead is moved to one side of the weld area, an undercut is produced on the shallow side of the bead. It may be necessary to grind the side walls of the weld area to minimize deep undercut conditions. If the root bead has a high, peaked center, it may also have to be removed with a grinder.

Another defect that can occur in the stringer, or root, bead is vermicular or wormhole porosity. The size of this porosity can vary over a wide range, but it is easily detected by radiographic inspection. Vermicular porosity occurs most often when welding high silicon content pipe. It is aggravated by excessive electrode travel speeds and high welding currents.

Welding cracks are more likely to occur in the higher-strength steels used for modern pipelines, such as X52 and X60. To prevent the occurrence of weld cracks, a number of factors must be controlled.

1. Joints must be prepared properly and root spacing must be correct.
2. High-low conditions must be minimized.
3. Welding must be done with electrodes designed for the grade of steel being welded.
4. Pipe may need to be preheated. The preheat temperature will depend on pipe strength, diameter, and wall thickness.
5. The line-up clamp should not be removed until the stringer bead is complete.
6. Pipe must be carefully lowered onto skids after the line-up clamp is removed.
7. Only enough welding current should be used to obtain a good bead, and travel should be slow.
8. Lack of penetration must be restricted.
9. Slag must be removed from each bead by power wire brushing.
10. Welders should weld stringer beads and hot pass with two or more welders on opposite sides of the pipe to equalize stress. Large-diameter pipe may require the use of three or four welders.
11. The hot pass should start within five minutes after completing the stringer (root) bead.
12. Wagon tracks should be minimized.

Other joining methods

Though most large, long-distance pipelines are constructed by welding individual joints together, other joining methods are used. Grooved couplings and threaded couplings are used for field flow lines, for example. Several joining methods are used for plastic pipe, including solvent cement, heat fusion, adhesive, and mechanical joints. Most of these other joining methods—and pipe materials other than steel—are used for special-purpose, short-distance, or small-diameter pipelines in which pressure and other operating conditions are not severe.

Considerable testing has been done, however, with a mechanical joint in gathering line construction.[8] The patented system involves cold-working both ends of each pipe joint to provide mating ends (Fig. 8–12). The bell end is expanded to a diameter slightly less than the outside diameter of the pipe. The other end, the pin end, has a groove rolled in the circumference with the end swaged slightly inward to allow it to start into the bell end. These ends are forced together using a portable hydraulic press, resulting in an interference fit. Epoxy resin is applied to the pin end to lubricate it during the joining operation. The resin then cures to form an O-ring seal in the groove.

Strength of the joint is obtained by the clamping action of the outer pipe (bell) on the inner pipe (pin) and depends on the clamping force and the friction between the two pipe ends. The clamping force depends on the amount of interference, the length of insertion, and the yield strength of the pipe steel.

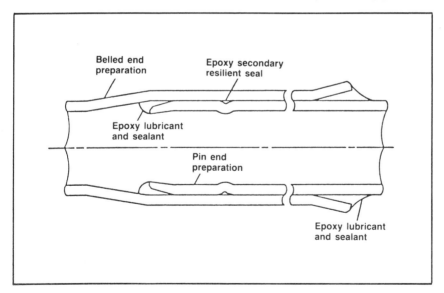

Fig. 8–12.   Mechanical pipe joint. Source: *Oil & Gas Journal,* 5 April 1982, p. 170.

A test program indicated that the insertion length required to obtain adequate joint strength ranged from 3¾ in. for 1½-in. and 2-in. nominal diameter pipe up to 13 in. for 12-in. nominal diameter pipe.[8] The required length is marked on the pin end prior to joining to ensure proper insertion.

Since the first pipelines were laid using this joint in the early 1970s, several improvements have been made in pipe end preparation, joining equipment, and procedures. Use of the method is said to have these advantages.

1. Requirement for welders is reduced.
2. A minimum of skilled personnel is required.
3. Pipeline construction is faster.
4. The method is less weather sensitive than welding.
5. X-ray requirements are reduced.

This method also has some disadvantages. The epoxy mix procedure is critical, joint inspection is required, and the special joining machine must be available.

Experience has indicated that it is not unusual to join pipe using this mechanical method at the rate of one joint/min. when a large number of bends is not required. With favorable conditions, a lay rate of 10,000 ft in 10 hr is not unusual. The method can permit the use of thinner wall pipe, reducing construction costs while still complying with applicable codes and regulations.

Other methods for joining pipe will likely be developed, but it is apparent that welding will continue to be used for most large, long-distance pipelines.

REFERENCES

1. *McGraw-Hill Encyclopedia of Engineering*. New York: McGraw-Hill Inc., 1982.
2. Bruno de Sivry, "Electron Beam Welding to Make J-Curve Pipelaying Possible," *Pipeline and Gas Journal*, (May 1982), p. 58.
3. "CRC Automatic Welding," Crutcher Resources Corp. (December 1982).
4. See reference 3 above.
5. United States Code of Federal Regulations, Title 49—Transportation; Subchapter D—Pipeline Safety.
6. Melvin A. Judah, and William L. Gloe, "Updating Federal Pipeline Safety Standards for Welding," *Oil & Gas Journal*, (10 May 1982), p. 138.
7. R.L. Looney, "Proper Welding Techniques Can Minimize Pipeline Defects," *Oil & Gas Journal*, (4 July 1977), p. 66.
8. W.L. Galey, "Patented Pipe-Joining Process Tested," *Oil & Gas Journal*, (5 April 1982), p. 170.

9

OPERATION AND CONTROL

EACH pipeline system has unique characteristics that dictate the type of control system that is most suitable. The complexity of control and operation ranges from a lease operator opening a valve to "run" a tank of oil into the gathering pipeline to sophisticated computer-based supervision that controls several hundred miles of pipeline, remote pump or compressor stations, and associated equipment from a central location.

Pipeline control systems can protect pipeline and equipment by monitoring and adjusting pressure and other operating variables, providing alarms when limits on operating conditions are exceeded, scheduling the shipment and delivery of different products, monitoring machinery performance and wear, controlling pressure surges in the pipeline, providing leak detection, and performing other functions. An individual system may not perform all of these functions, but these capabilities and more are possible with modern control systems.

The primary goal of a pipeline control system is to obtain the highest throughput at the lowest cost without exceeding pressure limits in the system and to deliver the required product volumes to the customer on schedule. Also, pumping, compression, and other equipment is monitored as discussed in detail in Chapter 11 to reduce the cost of maintenance. Today, predictive and preventive maintenance is an important approach to cutting operating costs.

Pipeline leak detection methods are also discussed in detail in Chapter 11. Leak detection is an important part of pipeline operation. Early detection of leaks can greatly reduce the loss of product from the pipeline and the danger of pollution.

Equipment monitoring and leak detection functions are independent of the supervisory control system. Additional computer analysis capability may be used for scheduling and other evaluation tasks.

Because of the variety of capabilities that exist in different pipeline control systems, only representative examples of such systems can be described here. However, these indicate what can be done in control system design and the tools that are used to operate oil and gas pipelines efficiently.

Supervisory control

Pipeline supervisory control systems regulate pressure and flow, start and stop pumps or compressors at stations along the line, and monitor the status of pumps, compressors, and valves. In a large pipeline system, many of the supervisory functions can be performed from a central location. The amount of control provided from a central location and the amount provided locally at individual pump or compressor stations varies widely. The control arrangement depends on the number and type of control functions required, the age of the control system, economics, and the preference of the pipeline operator.

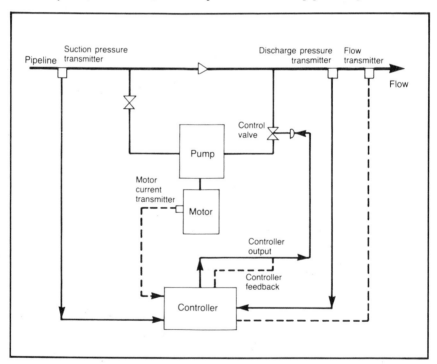

Fig. 9–1.   Typical pipeline control system. Source: *Oil & Gas Journal*, 30 August 1982, p. 126.

A typical basic pipeline control system is shown in Fig. 9–1. Figs. 9–2 and 9–3 show example control and dispatch facilities.

Fig. 9–2. Example pipeline system control room. Source: *Oil & Gas Journal*, 20 November 1978, p. 110.

Fig. 9–3. Pipeline system dispatch computer room. Source: *Oil & Gas Journal*, 25 October 1982, p. 171.

In general, modern computer-based pipeline control systems consist of the following elements:[1]

1. The computer complex includes computers, computer peripherals, and interfacing equipment for the man/machine system and remote stations.
2. The man/machine system includes devices necessary for the operator to communicate with the computer, such as the video display unit, keyboards, and loggers.
3. Remote stations are connected to the computer complex via a communications channel—microwave, telephone, radio, or other means.
4. Field devices, such as pumps and motor-operated valves, are controlled and monitored by the remote station. Field instrumentation includes pressure and temperature transmitters, tank gauges, and similar components.

Crude pipeline example.[2] A 747-mile, 48-in. crude oil pipeline in Saudi Arabia that includes 11 main line pumping stations and two pressure reduction stations is an example of a modern crude pipeline control system. The line is controlled and operated by a dispatcher located at the pipeline's western end at Yanbu. Each station's control house is linked to Yanbu by a microwave network; the station control house transfers commands to the operating units through control panels. The local control panel at each station acts as an intermediary during operation and as a main control panel for maintenance functions. Pump starting, stopping, loading, and unloading functions are performed automatically from Yanbu or the central control house.

Each pump driver is designed for unmanned operation with automatic self-protection, which is integrated into the overall supervisory control scheme. A separate monitoring system oversees major equipment, tells the operator the status of the machinery, and predicts future maintenance needs. Control of flow through the pipeline is done from Yanbu. Each station receives a set point command for either flow or flow equivalent (pressure increase across the station). When each pump unit reaches its set point, a confirming signal is sent to the dispatcher. Changes in the set point, the addition or deletion of main line pumps, and other changes can be made automatically from Yanbu.

Design flow rate can be met with two gas-turbine-driven pumps per station; the third unit is a standby. Lower flow can be handled by a combination of units, depending on volume requirements. At low flows, some stations may even be bypassed.

When a command is received at a pump station, it may be "conditioned" by local control equipment, then passed to the gas-turbine control house. Some commands—starting and stopping, for instance—do not require this conditioning. But set point commands that determine flow in the pipeline are integrated into local flow or pressure controllers. The condition and status of suction and discharge valves is always available to the controller at Yanbu. Local conditions or an interruption in the microwave communications link may cause the station control system to take direct control of units in the station.

A typical control sequence in this system is represented by a situation in which flow rate has been established and then a change in flow rate is required. At the established flow rate, one pump is operating at the station but another pump must be put on line to handle the increased flow rate.

The sequence in making this change is outlined as follows:

1. When the dispatcher initiates one of the two remaining gas-turbine-driven pumps, the start signal is transmitted to each unit sequencer via the station central control house and the unit remote panel. The sequencer initiates the opening of the suction, discharge, and bypass valves.
2. When suction and discharge valves are 60% open, the bypass is fully open, and all other station start conditions have been satisfied, air is fed into the starter and the turbine compressors begin to rotate. The power turbine is aerodynamically coupled to the gas generator, so no pump rotation occurs until energy (hot gas) is fed into the power turbine.
3. Suction, discharge, and bypass valves must be fully open before fuel is allowed to enter the unit and ignition is initiated. The gas generator starts to accelerate. When the turbine has reached self-sustaining speed and the pump has started to rotate, the sequencer will accelerate the turbine to the speed dictated by the set point request. It will also begin closing the bypass valve.
4. Since in this example another pump is operating in parallel with the pump being started, the speed of the operating pump will be automatically adjusted until the combined output of both units provides the desired flow rate.

Products pipeline examples. A French products pipeline serves as one example of how minicomputers can be used to monitor and control a complex system involving the transportation of several products to many delivery points. The pipeline network consists of 26 pumping stations and 34 delivery points. Total length is about 756 miles.[3] In 1979, the system transported about 158 million bbl of crude, motor fuel, domestic fuel, jet fuel, and naphtha.

Minicomputer-based control performs two main functions: the first is remote monitoring to acquire pressure, temperature, flow rate, alarm, and other data; the second is a data-processing function that checks the validity of data received, displays it on a video screen, controls the network, and measures the volume of a product delivered to a customer for billing.

In another example, the pump/driver control system for a natural gas liquids (NGL) pipeline combines analog and digital circuitry.[4] The electronic control system for the pipeline's 10 pump stations controls 34 turbine prime movers. It consists of 125 process controllers and 28 supporting transmitters for suction, discharge, gas production, and combustion process data. Additional equipment includes interrogators for signal selection, recorders, lightning-protection devices for signal transmitters, indicators for ignition and power-turbine temperature, and automatic/manual transfer stations. The transfer stations provide a bypass to keep the system operating while an engineer manually corrects problems in a controller or control loop.

One of the United States' large petroleum products pipeline systems delivers about 500,000 b/d over an average distance of 520 miles to terminals in the southeastern and mid-Atlantic states. It includes 23 pipeline segments totaling over 3,000 miles and ranging in diameter from 6 in. to 30 in. About 60 batches/day for a total of 39 individual shippers involving 131 different product entities must be monitored.[5] The system has 17 input points and 45 delivery points.

Before sophisticated computers were installed to operate the system, several generations of "hard-wired" equipment from different manufacturers were used in different sections of the system. The modern control and monitoring system (Fig. 9–4) now allows the pipeline operator to operate the system with as few as 16 employees—including two dispatchers—per shift.

The system installed on this products pipeline network contains some 20,000 status points that must be scanned every several seconds. The operator decided that one central computer could not do this, and the system was separated into computer hubs, each of which could be monitored by one large minicomputer. Control and monitoring data from the hub computers are retransmitted to a large master computer at the operations control center. This master computer is used for overall system surveillance and as a gathering point for measurement and power consumption data used in the power optimization program. Among the system's features are control preconditioning, a comparison of tank-level and metered volume, and automatic meter proving.

In preconditioning, the hub computer allows the operator to prepare in advance as many as five control functions in a sequence; each is triggered when the previous event in the sequence is completed. For instance, it is possible to prepare a sequence in which a tank is pumped empty, suction is immediately taken from a different tank, fluorescent dye is injected to mark the beginning of the new batch, a volume ticket is printed, and one pump is started and another pump stopped. The sequence can be prepared hours ahead of the time it will be initiated. When the tank signals "empty," the computer performs all the functions in the sequence in proper order.

The tank level vs metered volume comparison is done by the computer whenever a tank is active. If a discrepancy occurs, an alarm notifies operating personnel that an instrument malfunction has occurred.

The automatic meter proving feature provides the hub computer with the capability to correct turbine meter volumes for temperature, pressure, and

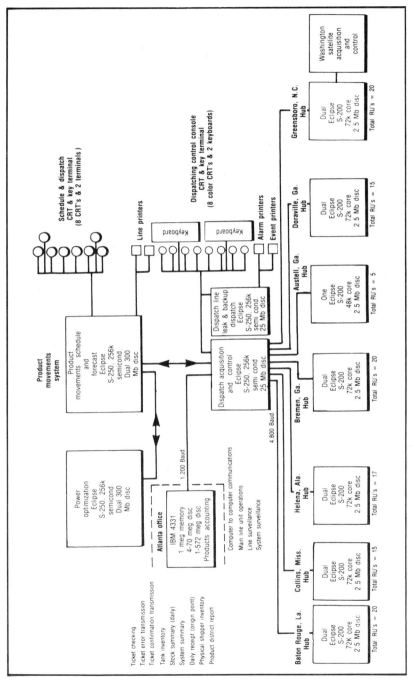

Fig. 9–4. Data gathering, monitoring, and control system. Source: *Oil & Gas Journal*, 25 October 1982, p. 171.

specific gravity once every 100 bbl. At least once during each batch, the meter is proved three consecutive times with a dedicated in-line meter prover. This is accomplished by a single command from the hub operator. When the three provings are accepted, the proper factors are calculated and applied to gross volumes. At the end of the batch, the net-barrel ticket is calculated by the hub computer and the shipper is notified of the net volume delivered.

Operating status reports (noncustody-transfer volumes) are also calculated by the hub computers for all input and output points.

A program was also designed for this system that would provide real-time computer-modeled leak surveillance. To be contained in the master backup computer, the program would survey every pressure, volume reading, and product characteristic from all seven hubs every few seconds and use the data in a complex model of every individual pipeline segment to provide highly sensitive leak detection. After testing on one segment of the pipeline, this capability was extended to the entire system.

Also planned is a power optimization program. Using hydraulic simulations, pump characteristics, and power company rate schedules, this program will select the most efficient pumping units to achieve a given pumping rate.

Gas pipeline system control.   Control of natural gas pipelines has much in common with liquids pipeline control. Many of the control functions are similar.

In addition to controlling flow, pressure, valve action, and other operation variables, the use of microprocessors and related equipment has seen increasing application to such services as optimizing engine/compressor unit efficiency and lowering fuel consumption. Microprocessor-based performance controls reportedly can reduce fuel consumption of large gas engines by real-time control of torque, air-fuel ratio, ignition timing, and the starting and stopping of units.[6]

Engine/compressor data can be provided by several means. Of significance is a comparison of the energy entering the unit in the form of fuel and the energy added to the gas by the compressor. In one application the processor calculates this ratio, an indicator of the relative efficiency of the engine/compressor unit. Monitoring the various readouts gives immediate information on the condition of the unit, operating parameters can be monitored for abnormalities, and long-term trends in the data can be analyzed to detect equipment deterioration and plan maintenance.

In this system, 10 two-cycle engine/compressor units with engine-driven combustion air blowers in pipeline transmission service are controlled. Fuel savings resulting from controlling maximum torque were as expected; an unexpected benefit was the large amount of data available instantaneously on a unit at one location. Though fuel savings by torque control and ignition timing were substantiated, savings by air-fuel ratio control were not as high as expected.

Microprocessors have also been applied to pipeline pressure controllers to reduce the consumption of valuable natural gas traditionally used for this purpose.[7] Pneumatic pressure control equipment at natural gas pipeline compressor stations typically uses gas from the pipeline as the power source. In a new approach, a microprocessor-based system can control pressure through electrohydraulically actuated valves with numerical solenoids. In this concept, fluid under pressure (air or hydraulic oil) is supplied by a pump to a pair of directional control three-way solenoid valves. When deenergized, both valves apply pressure to the cylinders of the valve actuator, locking the position of the control valve. When either directional valve is energized, fluid vents from the appropriate side of the actuator piston, moving the connecting rod that operates the valve. Fluid leaving the actuator cylinder returns to a reservoir. Selectively energizing the solenoids provides several different valve actuator motions. The control system can be used with a wide range of valve sizes and valve actuators by selecting the orifice size that will provide the proper fluid flow restriction.

Key features of this control method include the following:

1. Downstream or upstream pressure can be maintained at the value selected.
2. Direct position control allows a valve to be moved to any angular position between 0° and 90°.
3. Position override is possible for situations in which two pipelines are connected and it is desirable to have one valve follow the position of another valve that is actually controlling pressure.
4. Overpressure limit control permits the system to be programmed to shut down the pipeline in case of a sudden rise in line pressure above a preset limit. The valve will automatically reopen when pressure again falls to the set point pressure. The highest pressure in the line is stored for display to the operator.
5. For line break detection, the unit can be armed to detect sudden losses in pipeline pressure that may indicate a break in the line and the venting of gas.

Automation of pipelines and pump or compressor equipment is not limited to large long-distance transportation systems. Increasing fuel prices and labor costs have made automation justifiable in many smaller applications. In one project, for example, five satellite compressors used to move low-pressure casinghead gas from the field to a processing plant were automated.[8] Goals of the project were to provide a reliable alarm system to indicate when a compressor or dehydrator unit had shut down and to maximize gas throughput by continuously controlling first-stage suction pressure and engine rpm. Before automation, each of the three compressor stations had to be visited at least four times daily to check for malfunctions and to adjust compressor and engine to varying load conditions. Since the compressors were unattended for 16 hr/day, they were left with conservative adjustments during this time to minimize the possibility of a shutdown.

With the new control system, a computer requests data from data end devices that are passed to the computer by way of the remote terminal unit (RTU) and the master terminal unit (MTU). End devices feeding the remote terminal measure pressure, temperature, and engine rpm and send the data from the RTU to the MTU, which serves as an interface with the computer.

Data received by the computer from a station include field pressure, discharge pressure, and meter differential pressure. Alarms include engine down, high combustible gas, fire, low glycol dehydrator temperature, low instrument air pressure, and low field pressure. Also collected are engine rpm, engine manifold pressure and temperature, engine jacket water temperature, first-stage suction pressure, and second-stage suction pressure.

From these data, the computer can calculate the following:

1. Station throughput, using differential and static pressures
2. Brake horsepower, using engine manifold temperature and pressure, and a correlation supplied by the engine manufacturer
3. Real gas horsepower, calculated on a gas volume basis
4. First and second-stage rod compression and tension loads

The aim of the control program is to optimize throughput by maintaining the highest possible engine rpm and first-stage suction pressure. Increasing engine rpm or first-stage suction pressure is done only after several constraints have been satisfied. These include rod load, which must stay below a preset maximum; horsepower, which must not exceed a maximum; and engine rpm, which must remain between both maximum and minimum limits. A minimum pressure drop must be maintained between field and suction pressure, and engine jacket water temperature must remain below a preset maximum. When the highest possible suction pressure and rpm are reached, the control program continues to check variables and make appropriate adjustments.

An evaluation of this system indicates it has decreased compressor downtime and increased throughput capability.

In this—as in all pipeline control systems—maintenance is a key to reliability. A monthly preventive maintenance program requires inspection and calibration of each end device and remote terminal unit to ensure the data fed to the computer are accurate. Records on each device will indicate how far it was out of calibration at each inspection interval.

Control system reliability. Supervisory control systems for pipelines require maintenance. Personnel must be trained properly, and the effects of component failure must be considered in the design phase. Each of these elements is important in ensuring system reliability.

Redundancy is a common approach to increasing pipeline control reliability. It is the use of a duplicate of one element of the system so that if the primary unit fails the redundant unit can perform the tasks of the failed unit. In most redundant systems, if one element fails, its backup unit takes over automatically without the operator's intervention and with little or no upset of pipeline operating conditions.

Redundancy is useful in increasing reliability, but it must be carefully designed after considering the effects of failure of various control components.

Most pipeline companies repair malfunctioning control elements by replacing modules or components and sending the faulty element to the manufacturer for repair. Many recommend that maintenance contracts offered by computer manufacturers be used. This service can provide on-site maintenance and repair, eliminating the need for a large stock of expensive computer spare parts. Another approach is to provide a total unit replacement as a spare item. The malfunctioning unit is then returned to the manufacturer for repair.

Pressure surges. Pressure surges in pipelines are caused by shutting down pump stations or individual pumps, opening or closing valves, or the arrival at the pump station of an interface between two fluids. Control of these surges is important in operating a pipeline safely and at maximum efficiency.

A pressure surge is any change in pressure in the pipeline with no set limits of magnitude or rate of change of pressure.[9] Pressure surges travel through the liquid in the pipeline at sonic velocity, which varies from 3,000 to 4,000 ft/sec in most pipelines, depending on the type of liquid and the diameter and thickness of the pipe. One study of pressure surges recorded surges varying from a few tenths of a psi/sec to 2,600 psi/sec. Typically, however, the shutdown of a station causes an initial rate of change of pressure of about 150 psi/sec. The important factor in pressure surge control is the rate of change of pressure rather than the magnitude of the pressure.

Analysis of surge pressures in a pipeline system is complex, usually requiring the use of large computers and sophisticated programs. Modern hand-held calculators (programmable calculators), however, can be used to solve some of these problems.

The main reason for surge analysis is to determine if the pressure in the pipeline exceeds the maximum allowable transient operating pressure at any point in the line. This maximum allowable transient operating pressure is typically specified as 110% of the pipeline's design pressure, or maximum allowable steady-state operating pressure.[10] If transient, or surge, pressure is excessive, action must be taken to reduce its magnitude.

As an example of surge pressure action, assume a system includes a segment of pipeline with a centrifugal pump and check valve at the upstream end and a block valve at the downstream end. In the steady-state condition, flow rate and pressure gradient are constant. If the block valve is closed, however, flow rate at the valve decreases; pressure increases because kinetic energy is converted to potential, or pressure, energy. The surge pressure is the amount by which the pressure exceeds the steady-state pressure.

The surge pressure is propagated upstream until it reaches the pump, which responds to the change in pressure according to the characteristics of its head vs flow rate curve. As the pressure wave moves upstream, its characteristics are changed by friction in the pipe and by the elasticity of the liquid and the pipe. After reaching the pump, waves are also reflected back down the pipeline. Original waves and reflected waves can reinforce each other.

Pump discharge pressure must be properly controlled to avoid excessive line pressure that could cause rupture. A check valve on the discharge side of the pump is often used to protect the pump from backflow.

Scheduling

One of the most important functions of the pipeline operator—especially in the case of liquids pipelines—is to schedule the volumes of each product transported by the pipeline to ensure delivery to the customer at the desired time. This is a simple matter when product changes are infrequent. A complex system serving a number of customers with different products, however, requires frequent changes in flow and operating conditions.

Scheduling can involve complex, repetitive calculations by pipeline operations personnel. But this scheduling can be automated by the use of a computer, relieving operating personnel to better manage the pipeline system. One such automated scheduling system has reduced a two-day scheduling process to a matter of minutes.[11] The assignment of shipment and delivery time is done quickly by a computer, which also calculates the hydraulic rates at which the product moves through the pipeline. Hydraulic rates are a function of pump configuration, product mix, and line characteristics. The scheduling system will examine all possible pump configurations and choose that which moves the production in the desired time while minimizing the cost of power. In this application, a computer system controls the firm's pipelines and a backup computer acts as the process control unit for the pipeline.

A system such as this can be designed so that computer-support personnel are not required for routine use. The program in this installation, for example, were designed to be similar to the steps a schedular would perform if he were making the calculations manually. But the number of alternatives the computer can consider in a short time is much larger than is possible when manually calculating schedules. In this application, calculating a schedule consists of six steps:

1. Create or modify batch codes and volumes
2. Verify batch codes and volumes
3. Calculate starting conditions
4. Generate rate profile
5. Calculate shipment times
6. Calculate delivery times

Using the program, a scheduler can change volumes, add batches, delete batches, or move batches to a different sequence position. He typically makes an adjustment as soon as he is aware of the change by entering the change directly into the computer. The scheduler can reproduce the latest sequence and volumes on the computer terminal or he can produce a printed copy of the latest information.

This scheduling operation is similar to other systems, but two features of this system are reportedly unique: the ability to calculate a hydraulic rate profile quickly based on product, line, and pump characteristics; and the ability to examine all pump configurations and select the one that minimizes the cost of power.[12]

To calculate hydraulic rates for a given segment of the system, the computer searches for the maximum flow rate that satisfies all minimum and maximum suction pressure and maximum discharge pressures. Pressure loss caused by friction in the pipeline and changes in elevation is calculated, and suction and discharge pressure are determined for the specific product at the given rate.

To optimize power use, the cost of power during the time the product is transported at a constant rate is calculated. The amount of power used is determined from pump and motor efficiency curves applicable to those specific units. This power use, in kilowatts, multiplied by the time the rate is used is the number of kilowatt-hours required. After all of the product has been transported, the power cost is calculated using kilowatt cost, kilowatt-hour cost, fuel adjustment cost, and power facilities cost.

Reports generated by the system include a station report, a line report, and a tankage report. The station report contains all significant events organized by station in time order. The line report merges information contained in all station reports. A tankage report shows product levels by time period.

The time required for manual scheduling was compared with the time required for computer performance of the function. In this system, an evaluation showed time to calculate a schedule was 12 hours manually, five minutes by computer; to recalculate a schedule took about four hours manually but takes only five minutes by computer. The time required to calculate hydraulic rates and optimize pump selection takes the computer system about one hour for 175 pump configurations; it was not considered feasible manually.

Interface detection. One of the important operating functions in liquids pipelines, especially those in which more than one product is shipped, is to record the passage of the interface between two products, or *batches*. This capability is necessary to provide accurate measurement of the volume of each product shipped.

In products pipelines, for example, the difference in fluid properties between two products may be small and interface detectors must be quite sensitive. Densitometers have been used widely for interface detection, but the use of the sonic pipeline interface detector (Figs. 9–5 and 9–6) is gaining acceptance.[13] Sonic interface detectors precisely measure the velocity at which ultrasonic pulses travel over a liquid path of known dimension. Sound velocity is a property unique to each material, as is viscosity and density. Because the sound transmission characteristics of each liquid are unique, its passage can be detected by the sonic device.

There is a general relationship between specific gravity and sound velocity for petroleum products. Each product flowing in a pipeline can be represented by a range of sound velocities. The operating range of the detector can be adjusted to cover any group of products for which interfaces must be monitored. The devices are extremely sensitive; they can detect the interface between a low-lead regular gasoline and a premium gasoline, for example, even though the specific gravities of the two materials are very nearly equal. Detecting the interface between gasoline and naphtha, or naphtha and fuel oil, requires the detector be set on a broader operating range because of the larger difference between the specific gravities of the products.

Sound velocity is a function of the temperature and pressure at which the fluid is flowing, in addition to the composition of the fluid. When the sonic detector is used to distinguish between two products with specific gravities that are very nearly equal, temperature and pressure compensation is necessary to prevent the detector from mistaking a temperature or pressure change for a product interface. Temperature and pressure compensation are also required, however, when using densitometers under similar conditions.

Improvements continue to be made in these devices, and microprocessors have been combined with the instrument to obtain better control of variables and more reliable performance. Several hundred instruments were reportedly in service around the world in the early 1980s.[14]

Pigging

Pipeline pigs and spheres are used for a variety of purposes in both liquids and natural gas pipelines. Pigs and spheres are forced through the pipeline by the pressure of the flowing fluid. A pig usually consists of a steel body with rubber or plastic cups attached to seal against the inside of the pipeline and to allow

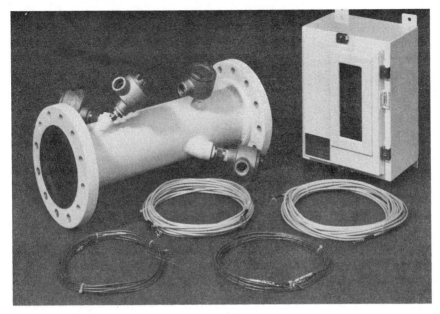

Fig. 9–5.   Sonic interface detector system. Source: *Oil & Gas Journal*, 30 November 1981, p. 80.

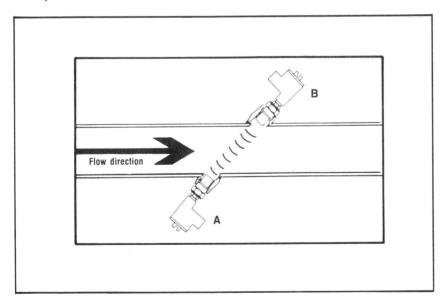

Fig. 9–6.   Transducer arrangement for interface detector. Source: *Oil & Gas Journal*, 30 November 1981, p. 80.

pressure to move the pig along the pipeline (Fig. 9–7). Different types of brushes and scrapers can be attached to the body of the pig for cleaning or to perform other functions.

Spheres are normally used to separate one fluid from another in a pipeline, either during hydrostatic testing of the line or during operation.

Pipeline pigging is done for the following reasons:[15]

1. To periodically remove wax, dirt, and water from the pipeline
2. To separate products to reduce the amount of interface between different types of crude oil or refined products
3. To control liquids in a pipeline, including two-phase pipelines, when filling lines for hydrostatic testing, dewatering following hydrostatic testing, and drying and purging operations
4. To inspect pipelines for defects such as dents, buckles, or corrosion using gauging pigs and electronic or caliper pigs
5. To apply internal coating to the walls of the pipeline for corrosion protection

Differential pressure required to move a pig or sphere through the pipeline overcomes the friction of the pig with the inside wall of the pipe. The force required depends on elevation changes in the pipeline, friction between the pig and the pipe wall, and the amount of lubrication available in the line. A dry gas pipeline provides less lubrication than a crude oil pipeline, for example.

Cups are designed to seal against the wall by making them 1/16 to 1/8 in. larger than the inside diameter of the pipe. As the cups become worn, the amount of *blow-by* (fluid bypassing the pig) increases because the seal is not as effective. In the case of spheres, the amount of inflation will depend on the purpose of the sphere. Pressure inside the sphere expands it against the inside of the pipe to provide a seal. In two-phase pipelines, spheres are sometimes underinflated to allow some blow-by to lower the density of the fluid ahead of the sphere.

Pigs and spheres travel at about the same velocity as the fluid in the pipeline. In liquid pipelines, the travel speed is relatively constant; in gas pipelines, however, the pig may travel awhile, then stop. Since the force required to start the pig is greater than the force required to sustain travel, a pig will continue travelling at a lower differential pressure than that needed to start it moving. Typically, a pig will stop at a circumferential weld in the pipeline.

Example of pigging operations. Pigging of gas transmission pipelines is done primarily to maintain efficiency by cleaning the pipe. Downstream of compressor stations, pipeline sections may require periodic pigging to remove lubricating oil from compressors. It can collect in low places in the pipeline and

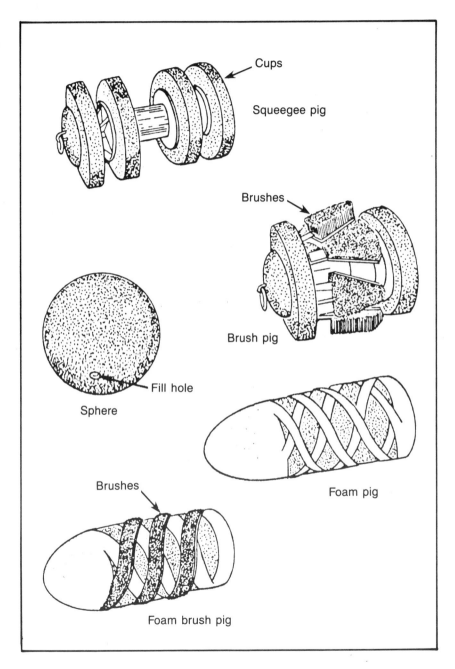

Fig. 9–7.   Types of pipeline pigs. Source: *Oil & Gas Journal,* 13 November 1978, p. 196.

restrict flow. Also, slugs of liquids inadvertently injected into the pipeline will collect at low elevations and reduce flow efficiency.

Cleaning pigs are used in all types of pipelines to increase efficiency and avoid problems at pump or compressor stations that could result from the presence of unwanted materials. Brushes on a cleaning pig remove dirt and wax from the pipeline walls. Several runs with the pig may be needed to adequately clean a section of the pipeline. Brush or scraper pigs contain holes that allow fluid to bypass the pig to prevent buildup in front of the pig that could cause plugging.

Very large amounts of debris can be removed by a pig if it is run over a long distance. For example, assume a pig is run in a 24-in. pipeline 100 miles long and removes 0.016 in. of wax material from the wall of the pipeline. After 100 miles, a plug about 1,450 ft long would form.[14]

Pipelines are often pigged first during testing following construction. Most pipelines are tested with water (hydrostatic testing) either in sections or over the entire length. A pig is normally sent ahead of the water when filling the test section to prevent mixing the test water with air in the line. This pig can also be instrumented to locate dents and buckles in the line. Internally coated pipelines are often flushed with water ahead of a pig to prevent debris from being dragged along the inside surface, damaging the coating.

After testing, the water is usually displaced with the fluid to be transported in the pipeline. A pig is run between the two fluids to separate them. In gas pipelines, the pig is used to "dewater" the pipeline by running it behind the test water. Additional pigs may also be run to ensure that as much moisture as possible is removed from the line.

Care must be taken when pigging pipelines to avoid the movement of foreign material into compressors and compressor-station piping and into pumping systems.

Special-purpose pigs have also been built (Fig. 9–8).

Launching and receiving.[16]   Equipment is required to introduce the pig into the pipeline and to retrieve the pig at the end of the segment being pigged. A launcher is required at the upstream of the section and a receiver at the downstream end. The distance between these pig "traps" depends on the service, location of pump or compressor stations, operating procedures, and the material used in the pig.

The amount of lubrication is a key factor in determining the distance between launching and receiving facilities. In gas transmission service, the maximum distance between traps has been recommended as 100 miles for pigs and 200 miles for spheres. In crude oil pipeline systems, the recommended distance between traps is 300 miles for pigs and 500 miles for spheres. These

Fig. 9–8.   A pipeline pig. Source: *Oil & Gas Journal*, 25 June 1979, p. 129.

distances represent extremes; the proper distance depends on the amount of sand, wax, and other material that will be carried along with the pig.

The design of pig launchers, pig traps, and related equipment is done in accordance with standards developed by several organizations. Traps for brush pigs, squeegees, and foam pigs include a barrel, short pup joint, a trap valve, a side valve, and a bypass line (Figs. 9–9 and 9–10). The barrel holds the pig for loading and unloading and is equipped with a quick-opening closure or blind flange. A barrel diameter 2 in. larger than the diameter of the pipeline served has been recommended. In large-diameter gas pipelines, the barrel diameter can be 1 in. larger than the pipeline. Barrel length depends on operating procedures, service, and available space.

Sphere launchers often must be designed for launching multiple spheres, so the barrels for sphere launchers are typically longer than those for other types of pigs. The operator can load these "magazines" with several spheres that can be launched automatically. This approach is often used in two-phase pipelines. The sphere launcher consists of the barrel, a launching mechanism, an isolation valve, an equalizer valve, and a reducing tee. A drain can serve as an equalizing line. Diameter of the launching and receiving barrels for spheres is typically 2 in. larger than the pipeline, and they can hold up to 10 or more spheres.

Combination pig and sphere launchers can also be designed if both cleaning pigs and spheres for liquid control are needed.

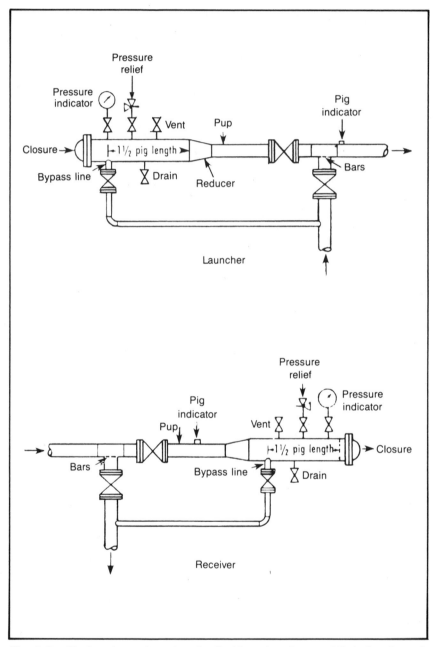

Fig. 9–9. Pig launcher and receiver for liquid service. Source: *Oil & Gas Journal,* 27 November 1978, p. 74.

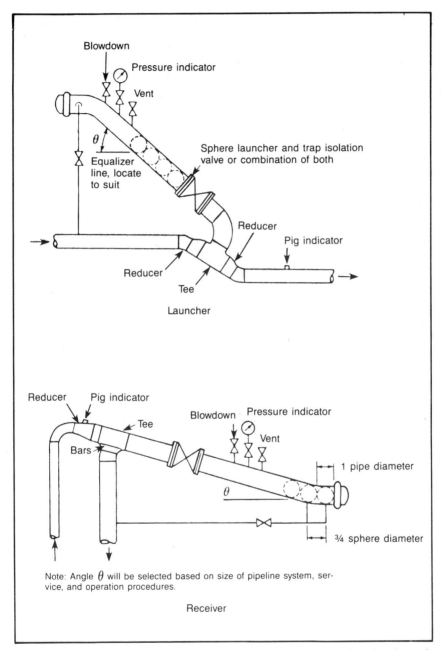

Fig. 9–10.   Sphere launcher and receiver for gas service. Source: *Oil & Gas Journal,*
27 November 1978, p. 74.

REFERENCES

1. R.E. Daves, "Improving Overall Reliability of Pipeline Supervisory Control System," *Oil & Gas Journal,* (14 June 1982), p. 68.
2. Chester Stasiowski, "Control System for Saudi Arabian Crude Line," *Oil & Gas Journal,* (28 December 1981), p. 201.
3. Bernard Vergnes, "Minicomputers Operate French Products Pipeline," *Oil & Gas Journal,* (8 February 1981), p. 130.
4. James D. Lewis and Elke R. Eastman, "Annual Pipeline Number: Electronics Control Pump Station Drivers," *Oil & Gas Journal,* (24 November 1980), p. 88.
5. Roger E. Guilford, "Plantation Pipe Line Takes Leap Forward in Automation of Products Line System," *Oil & Gas Journal,* (25 October 1982), p. 171.
6. Hans D. Lenz, "Microprocessors in Pipelining: Microprocessor-based Optimizing Systems Boost Engine/Compressor Unit Efficiency," *Oil & Gas Journal,* (14 December 1981), p. 151.
7. Jake L. Harris, "Microprocessors in Pipelining: Microprocessors Control Pipeline Pressure," *Oil & Gas Journal,* (14 December 1981), p. 154.
8. Bill Cagle, "Automation of Satellite Compressor Proves Valuable," *Oil & Gas Journal,* (28 January 1980), p. 155.
9. H.A. Brainerd, "Good Surge Control Can Help Pipeline Throughput," *Oil & Gas Journal,* (30 August 1982), p. 126.
10. Mike Hein, "Pipeline Surge Analysis—1: Calculator Can Ease Pipeline Surge Analysis," *Oil & Gas Journal,* (10 August 1981), p. 100.
11. Wayne Baxter, Anwar Gopalani, and Edmond Murray, "Pipeline Scheduling—1: Automated Scheduling System for Pipelines," *Oil & Gas Journal,* (28 September 1981), p. 323.
12. Wayne Baxter, Anwar Gopalani, and Edmond Murray, "Pipeline Scheduling—2: Automated System Minimizes Pipeline Power, Calculates Rates," *Oil & Gas Journal,* (5 October 1981), p. 148.
13. E.M. Zacharias Jr. and R. Ord Jr., "Developments Broaden Use of Sonic Pipeline Interface Detectors," *Oil & Gas Journal,* (30 November 1981), p. 80.
14. See reference 13 above.
15. Brian C. Webb, "Pipeline Pigging—1: Guidelines Set Out for Pipeline Pigging," *Oil & Gas Journal,* (13 November 1978), p. 106.
16. Brian C. Webb, "Pipeline Pigging—2: More Guidelines Given for Gas, Liquid Pipeline Pigging," *Oil & Gas Journal,* (27 November 1978), p. 74.

10

METERING AND STORAGE

MEASUREMENT of crude, natural gas, and natural gas liquids has always been an important part of pipeline system operation. But measurement accuracy has become much more important in the last decade because of the severalfold increase in oil and natural gas prices. The cost of inaccurate measurement and waste is now so great that an investment in sophisticated measurement equipment and the use of special techniques can be easily justified.

Ownership of petroleum products may change several times between the wellhead and the consumer. At each of these custody transfers, the buyer and seller want to be sure of the exact volume transferred so fair payment can be made. In both pipeline and tanker transportation, emphasis on accurately measuring volumes loaded onto and delivered from tankers and the prevention of losses has increased in recent years.

Accurate measurement is desirable even where custody is not transferred from one owner to another. In the field, the operator wants an accurate measurement of production from each well to help analyze well performance. Volumes transferred to and from storage must also be measured to avoid loss.

Volume is not the only variable important in measuring hydrocarbon streams. The value of natural gas depends in part on its heat, or energy, content. Energy content is often expressed in British Thermal Units (BTU) per standard cubic foot (scf). A natural gas whose heat content is 900 BTU/scf does not provide the consumer as much energy as one that has a heat content of 1,000 BTU/scf.

Traditionally, gas purchase contracts specified only a minimum BTU content, typically 1,000 BTU/scf. The seller had only to meet that minimum to receive the contract price for each standard cubic foot. He could remove valuable liquid components in the gas stream in a gas processing plant before selling the gas to the purchaser as long as the delivered gas had a heat content of at least 1,000 BTU/scf. Ethane, for instance, might be more valuable as a petrochemical feedstock than as a component in the natural gas stream.

When prices for natural gas liquids are low, removing them from the gas stream to sell as separate products may not be justified. If left in the sales-gas stream, they increase its volume. When these components are left in the gas stream, its heat content also increases. Of the individual components in natural gas, methane makes up the bulk; ethane, propane, and small amounts of heavier hydrocarbons may also be included. The gross heating value of methane is 1,009 BTU/cu ft; for ethane, it is 1,769 BTU/cu ft; and for propane, 2,517 BTU/cu ft. Greater amounts of heavier components in a gas stream increase its heat content.

When prices for ethane and propane are high relative to their value in the natural gas stream on a cubic-foot basis, more of these products are removed from the gas for other markets.

As the value of natural gas increased from 10–20¢/1,000 cu ft (Mcf) to $2–4/Mcf and more, emphasis was placed on measurement techniques that more accurately reflect the energy value of the gas. In the United States, regulations require that measurement of natural gas reflect its BTU content.

Measurement of crude oil also involves more than total volume. Crude oil usually contains entrained water and sediment (bottom sediment and water, or BS&W). Traditionally, the volume of sediment and water has been measured and the total volume passing the meter has been corrected when payment is made. This was an important part of crude oil measurement when crude sold for $3.00/bbl; it has become much more important as crude prices have climbed to $30/bbl and higher. When custody of a shipment of crude oil—either from a pipeline or a tanker—changes, the accurate measurement of water in the oil is critical.

Water content can be measured both manually and automatically. On the lease, a lease automatic custody transfer (LACT) unit contains pumping, metering, and BS&W measurement equipment. The unit automatically begins pumping from a lease storage tank into the crude gathering pipeline. When the pump has lowered the liquid in the lease tank to a prescribed level, the LACT unit shuts off automatically. Metering and BS&W measurement are done automatically. Determining BS&W content with automatic devices normally depends on measuring electrical characteristics of the stream. Crude and water have different electric resistance properties, allowing the detection of water in a crude oil stream.

Measurement of water in tanker cargoes of crude has also become critically important as crude prices have escalated. Accurate measurement of volume and water content are necessary to ensure that the buyer, for instance, is not paying $30/bbl for water rather than oil.

The importance of accurate measurement cannot be overstated, especially when ownership of oil, natural gas, or products changes. Custody transfer measurements are the basis of payment to the producer and the royalty owner and for the payment of taxes.

Orifice meters

One of the most versatile and widely used measuring devices is the orifice meter (Figs. 10–1 and 10–2). This instrument has been used for many years in oil and gas operations around the world. The flow of both gases and liquids can be measured with orifice meters; they are especially popular for natural gas measurement.

The meter itself is part of a meter station that includes the meter tube, a length of pipe upstream and downstream of the orifice; the orifice plate, which is installed vertically in the meter run; flanges on each side of the orifice plate that are tapped so pressure can be monitored; and a recorder. The recorder is clock-driven, and the pressure monitored at the orifice meter flange taps is recorded on a time-based circular chart. Some pressure-monitoring taps are located in the pipe rather than in the orifice meter flanges.

Measuring natural gas.   Natural gas may be measured with orifice, positive displacement, turbine, and other types of meters. These devices measure only the volume of gas flowing in the line. In recent years, emphasis on heat content

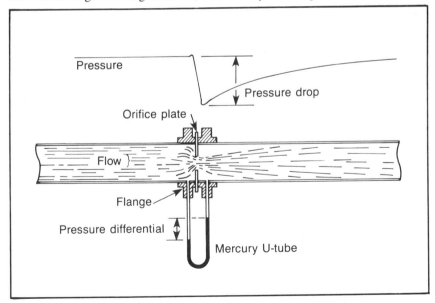

Fig. 10–1.   Orifice plate uses differential pressure to measure flow. Source: Reference 1.

Fig. 10–2. Orifice meter runs with block valves in foreground. Source: *Oil & Gas Journal*, 25 December 1978, p. 193.

of the gas has resulted in techniques for monitoring the BTU content of a flowing gas stream. In addition to the traditional methods that use periodic sampling or chromatography, for example, experiments have shown that acoustic measurement of gas BTU content may be practical.

Measurement of natural gas volume requires that the conditions at which the volume is determined must be stated. Measurements are normally adjusted to a base temperature and pressure, as discussed in Chapter 4. These base conditions normally are atmospheric pressure and 60 °F. The exact base conditions are spelled out in each gas purchase contract and vary slightly from area to area and from contract to contract. To calculate gas volume passing through a meter, other information is required in addition to base temperature and pressure. These data include flowing temperature and pressure, gas specific gravity, constants that have been determined for the specific meter, and the supercompressibility of the gas at flowing conditions.

Volume calculations are made using readings from the recorder chart and the appropriate correction factors. On sales-gas lines, where accuracy is especially important, the temperature of the flowing gas stream is continuously recorded on another chart so the proper temperature correction factor can be used in volume calculations.

By adding the flow volumes for each time increment on the pressure recording chart—integrating the chart—the volume passing through the line during a 24-hour period can be determined. The charts are collected, and a new chart is installed periodically.

Several types of pressure-monitoring devices are available for orifice meter stations. The type used will affect factors used in volume calculations. However, in all orifice meter installations, both the static pressure in the gas pipeline and the differential pressure across the orifice plate must be recorded to permit volume calculations.

The orifice plate is a key to measurement accuracy with an orifice meter. It is a round steel plate with a hole in its center that is inserted into the meter tube between the orifice flanges perpendicular to the axis of the meter tube. It restricts flow by reducing the diameter of the area through which gas can flow. Flow through this restriction reduces the pressure on the downstream side of the orifice plate. This pressure drop, measured through flange taps or pipe taps on each side of the plate, is the key value in volume calculations.

Orifice plates are manufactured according to specifications of the American Gas Association (AGA) and other organizations.

In flowing through an orifice plate, the potential energy (pressure energy) of the gas is changed to kinetic energy. The relationship between the potential energy upstream of the orifice plate and the kinetic energy downstream of the plate is the basis for calculating the volume of gas passing through the orifice. Flow through the plate is directly related to the static pressure in the line and the differential pressure caused by the restriction of the orifice.

Many other factors are required in accurately calculating the flow through an orifice meter. These factors have been calculated and set forth in tables of constants to be used when calculating flow volumes. If these several meter factors are combined into one constant, the quantity of gas measured by an orifice meter can be described:[1]

$$Q_h = C'h_wP_f$$

Where:

Q_h = rate of flow at base conditions, cu ft/hr
h_w = differential pressure, in. of water
P_f = static (gauge) pressure, psia
C' = orifice flow constant

The orifice flow constant, in turn, consists of several factors that can be obtained from published tables:

$$C' = F_b \times F_{pb} \times F_{tb} \times F_g \times F_{tf} \times F_r \times Y \times F_{pv} \times F_m \times F_l \times F_a$$

Where:

F_b = basic orifice flow factor
F_{pb} = pressure base factor

F_{tb} = temperature base factor
F_{tf} = flowing temperature factor
F_g = specific gravity factor
F_r = Reynolds number factor
Y = expansion factor
F_{pv} = supercompressibility factor
F_m = manometer factor
F_l = gauge location factor
F_a = orifice thermal expansion factor

Orifice meter run design and maintenance. The size of the orifice plate must be selected according to the diameter of the pipe used in the meter tube and the expected flow volume. The outside diameter of the plate and the size of the plate holder depend on the size of the meter tube; the size of the hole in the orifice plate depends on the expected flow. When flow through the pipeline and meter tube changes, the orifice plate can be changed without changing the meter tube. A given orifice size will measure a range of flow volumes, so the plate does not have to be replaced with each minor change in flow. But when significant changes in volume occur, the plate must be changed to provide accurate measurement.

It is usually recommended that the recorder operate in the upper half of the chart, for example, for most accurate results. If flow decreases significantly and the recorder operates near the axis of the circular chart, accuracy is diminished and an orifice plate with a smaller hole should be installed to bring the reading back into the desired area of the chart. Also, the ratio of the size of the orifice to the inside diameter of the meter run should be less than recommended maximum.

The meter tube—the section of pipe on each side of the orifice plate holder—is important in ensuring accurate gas measurement in an orifice meter station. It must be properly sized, manufactured, and installed. Manufacturing the tube to specifications and installing it properly are necessary to provide streamlined flow of gas through the tube and avoid turbulence.

The meter station should be placed where it is easily accessible for checking and changing recorder charts. The meter tube must be placed aboveground and connected to the buried pipeline. This connection is typically made by installing 45° ells in the buried pipeline, a short section of pipe at a 45° angle to the ground, then another 45° ell at each end of the meter tube at the desired height above the ground. The result is a horizontal meter tube. The 45° ells, rather than 90° ells, are recommended to avoid sharp changes in the direction of gas flow that could cause turbulence.

The upstream edge of the orifice plate's hole must be sharp and must not contain nicks or burrs that could cause turbulence as gas flows through the restriction. Some plates have holes that are bevelled from the upstream to the downstream side of the plate. The maximum angle of this bevel is also specified in the AGA guidelines, as is the thickness of the orifice plate. The orifice must also be concentric with the ID of the tube.

Periodic inspection and maintenance of orifice plates is required to ensure they have not become damaged. Solid particles in the gas stream may chip the upstream edge of the hole in the plate, for example, resulting in turbulence. Depending on the type of orifice plate holder, removal of the orifice plate may require that flow be stopped and the pressure in the meter run bled off. Most orifice plate holders, however, allow removal of the orifice plate without interrupting flow.

The inside surface of the meter tube should also be free of pits or other obstructions that could cause turbulence near the orifice plate. The tube's inside diameter must be within specified tolerances because it affects volume calculations.

Length of the upstream and downstream sections of the meter tube are also outlined in specifications to ensure that any turbulence created by valves, reducers, or other fittings on either side of the orifice plate has dissipated before reaching the orifice. Straightening vanes may also be used in the upstream section of the meter tube to eliminate turbulence ahead of the orifice plate.

In addition to ensuring that a meter run is designed and installed to avoid turbulence caused by pipe fittings and other obstructions, it is important that pulsation not be present in the flowing gas stream. Pulsation causes erroneous pressure readings because of the constant fluctuation of pressure in the line. These fluctuations on the recorder chart make it difficult to read the proper value from the chart, and accuracy is reduced. The main cause of pulsation is reciprocating compressors.

When pulsation is present, it must be eliminated before accurate flow measurement is possible or the meter run must be relocated to a position on the pipeline where pulsation does not exist. Pulsation dampeners can be used to reduce or eliminate pulsation, but their design and placement is complex and satisfactory results are often difficult to obtain.

Liquid measurement. Liquids and steam may also be measured with orifice meters by applying appropriate correction factors. The factors differ from those used when measuring natural gas.

The same general considerations apply to the design and use of orifice meters for fluid measurement. Opportunities for turbulence should be eliminated, the size of orifice and meter tube must be within specified tolerances, proper factors must be used, and periodic inspection and maintenance are required.

Positive displacement meters and turbine meters

Positive displacement meters and turbine meters can also be used to measure both liquids and gases. They are, however, more commonly used for liquids measurement in both pipeline and tanker loading operations.

In positive displacement meters, the fluid passes through the meter in successive isolated quantities by filling and emptying spaces of fixed volume. A counter registers the total quantity of fluid passing through the meter. Some meters of this type have a flow rate indicator in addition to a total flow recorder.

Turbine meters. The force of the flowing fluid turns a bladed rotor in a turbine meter. The axis of the rotor is parallel to the direction of flow, and the rotor's speed of rotation is proportional to the flow rate. Using proper gearing, the revolutions of the rotor are related to volume to provide a volume indication. Turbine meters are installed in a meter run, usually horizontally. Many of the considerations discussed earlier regarding orifice meter installations are also important in turbine meter runs.

In designing a turbine meter station, the expected flow range, flow conditions (continuous or intermittent), operating pressure, pressure drop, and temperature must be considered. Also important is the type of fluid, including its viscosity and corrosive properties and the presence of solids or water.[2] In gas service, care should be taken during startup to prevent overspeeding the rotor.

Turbine meters can measure high and low flow rates with the same meter, and the meters have high repeatability. Several types of readouts are available, including a digital readout and a ticket printer. An important part of any displacement or turbine meter installation is a meter prover, used to calibrate the meter. The prover provides the necessary correction factors for adjusting the "raw" readings to an accurate volume.

The volume flowing through a turbine meter is calculated by adjusting the uncompensated meter pulses to base conditions of pressure and temperature and allowing for the effects of flowing temperature and pressure on the fluid and on the meter. In one application of turbine meters in an automatic system, for instance, the meter throughput is determined as follows:[3]

Volume at reference conditions = (meter pulses/MKF) × CPLMR × CPSMR
                                  × CTLMR × CTSMR

Where:

MKF = meter K factor obtained from meter proving, in meter pulses/cu m

CPLMR = pressure correction factor for liquid in the meter to reference conditions

CPSMR = pressure correction factor for steel in the meter to reference
conditions
CTLMR = temperature correction factor for liquid in the meter to
reference conditions
CTSMR = temperature correction factor for steel in the meter to reference
conditions

In this installation, 12-in. turbine meters are part of an automatic system in which a computer counts the pulses from the meters and performs pressure and temperature compensation calculations to account for pressure and temperature effects on both the liquid and the meter housing. In this case, reference conditions are specified as 15°C (59°F) and 1.01325 bars absolute (14.9 psia).

This automatic computer turbine metering system provides complete control of metering and meter proving. It is a batch operation in which crude is loaded from an offshore platform's storage onto a crude tanker for shipment. The importance of measurement accuracy is obvious in this installation. The system can deliver about 68,000 bbl/hr through four of the five 12-in. turbine meters. At $30/bbl, this represents about $2 million worth of crude per hour. An error of 0.05% represents an uncertainty of about $800,000 per year based on 300 days of field production at a rate of about 180,000 b/d.[3]

Meter proving. A meter prover (Fig. 10–3) is used with positive displacement and turbine meters to establish a relationship between the number of revolutions, or counts, of the meter and the volume flowing through the meter. Meter provers are available in several types. They all are capable of very accurately measuring a specified volume of fluid and comparing the number of meter counts that corresponds to this volume. Provers typically consist of a length of pipe that contains detector switches to record the passage of a sphere. The volume between the two detector switches is accurately known. These switches are electronically connected to a counter that measures the time for the volume between the two spheres to pass between the two detectors. The number of counts on the meter being proved is related to the volume passing the detectors.

Additional equipment on a typical bidirectional, folded-type prover system (Fig. 10–3) includes manifolding and appropriate valves to direct the flow, piping to connect the prover with the meter run being tested, provisions for checking for leaks, and other equipment. This type of displacement prover must be maintained properly to prevent leakage across the spheres, leakage through valves, excessive wear or deterioration of the spheres or barrel lining, and improper operation of the detectors. The prover shown in Fig. 10–3 has enlarged sections at the ends of the barrel. To prevent deformation of the spheres, they are kept there when the prover is not in use.

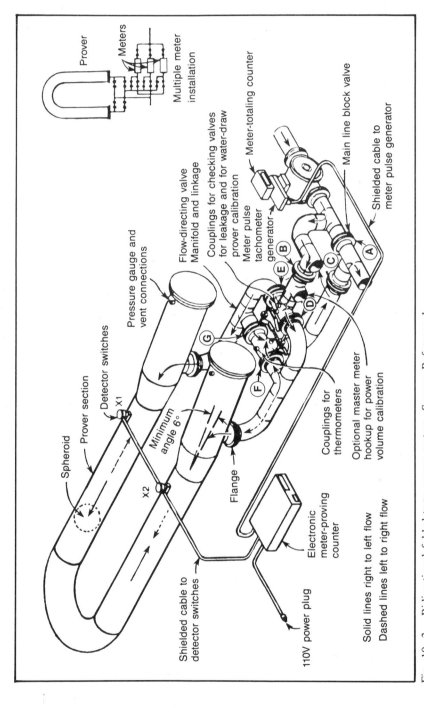

Fig. 10–3. Bidirectional folded-type prover system. Source: Reference 1.

Prover

Meters

Multiple meter
installation

Meter-totaling counter

Main line block valve

Shielded cable to
meter pulse generator

Flow-directing valve
Manifold and linkage

Couplings for checking valves
for leakage and for water-draw
prover calibration

Meter pulse
tachometer
generator

Pressure gauge and
vent connections

Detector switches

X1

Spheroid

Prover section

Minimum
angle 6°

X2

Flange

Couplings for
thermometers

Optional master meter
hookup for power
volume calibration

Shielded cable to
detector switches

Electronic
meter-proving
counter

110V power plug

Solid lines right to left flow
Dashed lines left to right flow

A

B

C

D

E

F

G

The metering installation described earlier includes a 36-in. bidirectional prover loop with a 16-in., 4-way diverter valve. The prover barrel is internally coated, and a water-inflated plastic sphere is used to sweep the volume between the detector switches. In this system, the operator instructs the computer to begin the automatic proving operation. The computer selects a meter to be proved, opens the prover inlet valve, and closes the meter outlet valve. The computer issues set-point outputs to the five flow controllers to balance all flow lines, so flow through the prover equals the operating flow rate of the meter.

The computer will conduct up to ten proving runs to establish a meter K factor. The K factor is considered acceptable when the average from five consecutive runs is within 0.05%. The meter factor is determined by comparing the volume indicated by the meter with the known volume of the prover.

If flow rate, temperature, or other operating conditions change, new meter factors must be determined. In the installation discussed here, the computer will alert the operator if limits set in the computer are exceeded. The operator then will initiate a proving operation to determine a new meter K factor.

Flow nozzles and venturis. There are other ways to measure the flow of fluids in addition to orifice meters, positive displacement meters, and turbine meters. Both the flow nozzle and the venturi meter are also differential pressure measuring devices, as is the orifice meter.

The flow nozzle is an elliptically shaped device inserted in a flow line that increases the velocity of the flowing fluid. Pressures at the entrance to the nozzle and its throat are measured to provide the differential pressure needed in flow calculations. Equations used with the flow nozzle are similar to those used to calculate flow through an orifice meter, with the substitution of appropriate correction factors.

The venturi is a tube consisting of a short, constricted portion between two tapered sections. The difference in pressure between the inlet and the throat of the device is proportional to flow through the tube. There is a variety of venturi tube geometries. Though general factors have been developed for use in flow calculations, it is usually necessary to calibrate each venturi to ensure accurate measurement.

Vortex meters. Another device used to measure gas, liquid, and steam is the vortex flow meter. It is a relatively new application of a phenomenon that was discovered many years ago.

Flow measurement using a vortex flow meter is based on vortex shedding that occurs when a fluid flows past an obstruction placed in the flow stream (Fig. 10–4). The fluid forms boundary layers on the surface of the obstruction that separate from the front corners of the obstruction and roll into vortices, or eddies, downstream. To cause vortices to develop, the obstruction must not be streamlined and is often referred to as a *bluff body.*

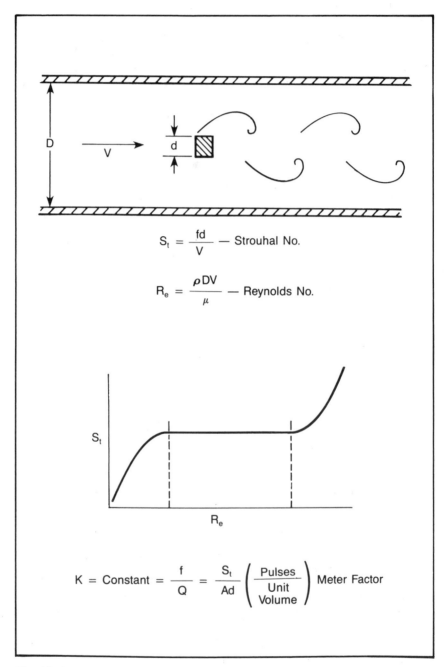

$$S_t = \frac{fd}{V} \quad - \text{ Strouhal No.}$$

$$R_e = \frac{\rho DV}{\mu} \quad - \text{ Reynolds No.}$$

$$K = \text{Constant} = \frac{f}{Q} = \frac{S_t}{Ad} \left(\frac{\text{Pulses}}{\frac{\text{Unit}}{\text{Volume}}} \right) \text{ Meter Factor}$$

Fig. 10-4.   Vortex flow meter theory. (Courtesy Fisher Controls).

Vortices are shed by the bluff body from one side, then from the other. The frequency of these vortices is proportional to flow velocity, allowing flow to be measured by monitoring the frequency of vortex formation. Design of the bluff body is important. The design must cause flow conditions in which the Reynolds number is well into the turbulent flow regime; otherwise, vortex shedding will not occur. Sharp corners improve the strength and regularity of the vortices. Some vortex meter devices contain two bluff bodies in the flow stream to reinforce the vortices and provide a stronger measurement signal.

Most designs are aimed at creating flow conditions that result in a Reynolds number ranging from 10,000 to 1,000,000. Over this range, vortex frequency is directly proportional to volumetric flow rate and is calculated by the following equation:[4]

$$Q = \frac{f(A \times d)}{(St)} = f/K$$

Where:

Q = volumetric flow rate
f = frequency of vortex formation
A = meter cross-sectional area
d = width of bluff body
St = Strouhal number, a dimensionless proportionality ratio that is constant over the Reynolds number range
K = meter K factor, pulses/unit volume

Flow calculations are similar to those involved in orifice meter flow calculations, but the equation is simpler than the orifice plate equation. There is no square root relationship, making the vortex meter capable of operating over a broader range. Also, no density relationship or viscosity correction is involved.

In a meter installation, sensors monitor the frequency of the vortices by measuring the pressure difference in the line caused by vortex shedding. Electronics amplify the signal, and the pulse frequency output is read from a meter. A filter can also be used in the circuit to reduce noise signals that may interfere with measurement.

Vortex flow meters had only a small share of the measurement market in the early 1980s. Some expected their use to grow much faster than differential pressure/orifice plate, turbine, and other meter types and to replace a significant number of existing differential pressure/orifice plate meters.

Mass measurement

In recent years, more emphasis has been put on mass flow measurement as an alternative to volume flow measurement. Mass flow measures flow in lb/hr, for example; volume flow measures flow in cu ft/hr. The two are related by the specific gravity, or density, of the flowing fluid.

Volume flow measurement requires that pressure, temperature, and composition of the fluid be used to express density. The key advantage of mass flow measurement is that density can be measured directly. This is particularly important when measuring fluids whose densities vary frequently as measurement conditions change. Mass measurement eliminates the need to define standard conditions. A pound of fluid remains the same regardless of temperature and pressure conditions. An example application of mass measurement in which this advantage is important is the measurement of natural gas liquids and other light hydrocarbon streams.

There are several ways to measure mass flow. Some meters measure flow directly by relating a constant torque or acceleration to mass flow; other systems use a volume flow meter in conjunction with a densitometer (density meter) to infer mass flow. The use of traditional volumetric measuring devices coupled with density measuring equipment is a common approach. In a mass measurement system, for example, a density meter might be used with a turbine meter (Fig. 10–5). The data from both devices can be integrated continuously with a minicomputer.

Both positive displacement flow meters and turbine meters have been applied to accurate mass measurement systems.[5] A meter prover system is required to determine the meter factor for either type of installation.

In a mixed stream of natural gas liquids, it is still necessary to know how much of each component is being delivered. A mass measurement system, in addition to a flow meter and density meter, also may contain a gas chromatograph and sampling equipment. A sample of the flowing fluid is taken at increments proportional to the flow rate, and analysis of the sample by gas chromatography provides the percent by component of the total stream.

Density measuring devices fall into two general categories: direct and inferential. The first provides direct weight measurement of a known volume of fluid; inferential density devices relate density to some physical law. The gravitometer is the most common method used in the measurement of natural gas liquids. Inferential-type devices used for this application include the vibrating vane and the vibrating tube meter, the buoyancy meter, and the radiation attenuation meter.

Operation of a mass-measurement system requires care, especially when custody transfer is involved. When measuring natural gas liquids, one report suggests these considerations:[6]

1. Single-phase liquid flow must be maintained.
2. Operating pressure must be carefully considered throughout the entire operating range since the fluid's equilibrium vapor pressure changes as composition and temperature vary.

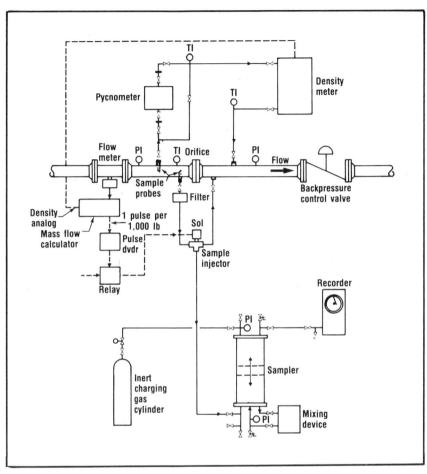

Fig. 10–5.   Mass measurement system. Source: *Oil & Gas Journal*, 27 April 1981, p. 270.

3. Filtering the stream is recommended if the system is subject to dirt problems. Dirty flow streams can cause accumulations on flow and density meters and result in inaccurate readings or failure.

4. Differences in operating pressure among the devices—flow meter, density meter, meter prover, etc.—can cause calibration errors.

5. Systems that use electronic digital integration by microprocessor or minicomputer should have integral procedures for verifying the accuracy of computing circuits.

6. The overall system should be proven frequently at operating conditions.

An example of a mass measuring system is a number of ethylene metering stations based on turbine meters, which are used for custody transfer. Ethylene density varies widely with temperature and pressure at normal pipeline conditions, and the mass measurement approach offers the best accuracy, according to a report on the system.[7]

The key to this system's accuracy is an insulated, bidirectional, piston-type prover of conventional configuration. The pipe is honed to a tolerance of ±10 mils and coated with a baked-on phenolic epoxy. A lightweight aluminum piston has spring-loaded, fluorocarbon seals to minimize sliding friction and provide resistance to chemical action of the ethylene. Proximity-type detector switches do not protrude into the prover barrel.

BTU measurement

The increased value of natural gas—and the fact that pipeline systems have become increasingly interconnected and the gas from several sources mixed— has brought a trend to the measurement of the heating value of natural gas rather than just its volume. Accurate measurement of heating value permits customers to be charged fairly and the producer to receive a fair price.

The common unit of heating value is the British Thermal Unit (BTU). There is still some variation in how the BTU is defined, and in gas purchase contracts it is wise to detail its definition precisely. Basically, however, it is the amount of heat required to increase the temperature of a given mass of water 1°F at a specified temperature. For instance, 1 BTU can be defined as the heat required to raise 1 lb (avoirdupois) of water 1°F from 58.5° to 59.5°F at standard pressure.

Heating value, or calorific value, should be further defined as either gross heating value or net heating value. Gross, or total, heating value of natural gas, for instance, is the "number of BTU evolved by the complete combustion at constant pressure of one standard cubic foot of gas with air; temperature of gas, air, and products of combustion being 60°F, all water formed by combustion reaction being condensed to liquid state."[8] Net heating value is the total or gross heating value minus the latent heat of vaporization at standard temperature of water formed by the combustion reaction.

These subtleties are not necesary, however, to understand the concept of heating value and its importance in natural gas measurement. In general, the heating value of a hydrocarbon increases with its molecular weight. For instance, methane, the primary component of natural gas, has a net heating value of 909 BTU/cu ft; ethane, 1,618 BTU/cu ft; and propane, 2,316 BTU/cu ft. Crude oils have much higher heating values because they contain large amounts of heavier hydrocarbons. The measurement of heating value for custody transfer purposes, however, is currently limited to natural gas.

Traditionally, heating value has been determined by batch methods, such as the chromatograph in which a sample is taken periodically from the gas stream and analyzed (Fig. 10–6). When the proportion of each component in the stream is determined, the heating value of each component, along with its percentage of the total, is used to determine the overall heating value of the gas.

Recent work has indicated the possibility of using acoustic measurements to monitor the BTU content of natural gas streams.[9] The speed at which sound travels and the BTU content of the components in a gas both depend on molecular weight. Relating sound speed to molecular weight therefore can provide BTU measurement. Based on natural gas mixture measurements made in one experiment, it is possible to detect, for instance, changes in the ethane/methane proportion on the order of 0.01%. This is comparable to the sensitivity of an analysis by gas chromatography.

Further development of the technique is required in order to be able to analyze the fraction of the gas stream consisting of noncombustible gases. Most gas streams contain small amounts of CO_2 (carbon dioxide) and N_2 (nitrogen), which do not contribute to the heating value of the gas.

Storage

Storage facilities for crude, natural gas, and light hydrocarbon products are an important element in all pipeline and tanker transportation systems. Storage

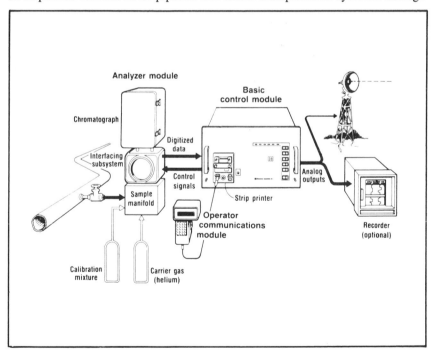

Fig. 10–6.   Heating value measurement system. Source: *Oil & Gas Journal*, 25 August 1980, p. 132.

allows flexibility in pipeline and refinery operations and minimizes unwanted fluctuations in pipeline throughput and product delivery.

The capacity of individual storage tanks or facilities varies widely. Small crude storage tanks are common on producing leases, while export tanker loading terminals can have several million bbl of crude storage capacity in a few giant tanks.

Both aboveground storage and belowground storage are used for both natural gas and hydrocarbon liquids.

Crude storage. On the producing lease, oil from individual wells is accumulated in tanks, then pumped into the crude oil gathering pipeline. Typically, one or more 500-bbl or 1,000-bbl tanks are located on an individual lease, depending on the volume of crude produced. Crude may not be pumped continuously from the lease tanks if produced volumes are small. If a LACT system is used, the tank will fill to a prescribed level and the pump will automatically start. When the fluid in the tank is lowered to another preset level, pumping into the gathering pipeline will stop automatically.

If the crude is run manually, an operator will start the pump when the tank is full and will stop it when the tank is nearly empty.

LACT units include meters to measure the amount of crude pumped. Meters are also installed in manually operated pumping systems. In some cases, when pumps are operated manually or crude is collected by truck, the tank is *gauged* before and after delivery to determine the volume delivered. Tanks are gauged by manually lowering a metal graduated tape with a weight bob on the end through a hatch in the top of the tank. Multiplying the depth of oil in the tank by a calculated factor given in bbl/ft of depth gives the volume in the tank. The volume gauged before delivery minus the volume after delivery is the amount of oil delivered.

Free water in the tank below the oil can also be determined by gauging. One technique is to coat the gauging tape with a material that changes color when contacted by water. Pumping is stopped a safe height above the water level to avoid pumping water into the gathering pipeline.

Other methods of determining the level in a storage tank are also used. A float on the inside of the tank connected to a graduated scale on the outside gives the fluid level at a glance. An automatic, remote gauging system can also be used. It provides a readout at a remote location of the liquid level in any tank connected to the system. It can be made to read directly in barrels or gallons when the appropriate mathematical conversions are used.

At each step in the pipeline gathering and delivery system, varying storage requirements exist. The point at which a gathering pipeline system enters a main crude trunk line, for example, often contains storage facilities. At the terminus of the crude trunk line—either a refinery or an export terminal—storage capacity is typically much larger. A refinery must operate continuously to be efficient, though its throughput can be reduced if necessary; enough crude to feed the refinery must be available. A crude trunk line is seldom shut down, but variations in throughput caused by conditions in the gathering system or in the producing fields are not unusual. Refinery storage minimizes the effect of these fluctuations and allows the refinery to operate at a relatively constant throughput.

A refinery located near oil production or at the end of a pipeline serving a major field may need less storage than a refinery that receives crude by tanker. Storage must be provided for periods between tanker deliveries. A refinery must also have storage for the products it manufactures.

Other factors that help determine the amount of storage a refinery needs for both feedstock and products include:[10]

1. *Product delivery.* If products are delivered continuously to a pipeline, storage needs will be reduced. Small refineries may accumulate products before delivery to the pipeline or to a tanker.
2. *Refinery complexity.* The more products a refinery produces, the more storage will be needed for intermediate feedstocks.
3. *Seasonal demand.* One product may be produced in greater quantities during a part of the year, another during another season. In the United States, gasoline sales are greater in the summer, for instance, and distillate fuel sales are greater in the winter.
4. *Turnarounds.* Refineries are periodically shut down for inspection, repair, and maintenance (a turnaround). Storage must be available for feedstock and products during these periods.

According to one source, a refinery that is supplied by pipeline should have 6 days of crude feedstock storage; if supplied by tanker, 32–35 days of storage is recommended.

At tanker terminals (Fig. 10–7), the amount of storage depends on the amount and type of product handled; the number, capacity, and type of tanker berths in the terminal; expected periods of downtime; and the number and size of ships using the terminal.

Crude storage tanks are cylindrical and are operated at near atmospheric pressure. Small lease storage tanks are typically shop-fabricated and are delivered to the site where they are connected to pumps and other facilities. Large crude storage tanks may be capable of storing up to several hundred thousand barrels each and must be built on the site. Large crude storage tanks often have a floating roof that moves up and down with the liquid level in the tank to minimize vapor losses. Smaller storage tanks, including those on the producing lease, have fixed roofs.

Fig. 10–7. Crude oil and product storage. Source: *Oil & Gas Journal,* 26 June 1978, p. 84.

Many crude storage tanks are equipped with vapor recovery systems that capture light hydrocarbons that evaporate from the crude and would otherwise be lost to the atmosphere. Economics, air-pollution regulations, or both may dictate the use of vapor recovery systems, depending on the area and the amount of vapor losses.

Considerations involved in designing large crude storage tanks include the integrity of the foundation, safety, maintenance requirements and ease of maintenance, and operating efficiency.

Underground caverns mined from salt domes or other formations are used for long-term crude storage. This type of storage is usually for strategic or emergency purposes. The United States is storing crude in its Strategic Petroleum Reserve in Texas and Louisiana, for instance, in order to be prepared for interruptions in the supply of imported oil. Crude is stored in five salt dome caverns; pipelines connect the caverns to crude terminals. Crude imported by tanker is pumped into the caverns through wells and will be withdrawn through wells when needed.

In early 1983, the Strategic Petroleum Reserve caverns held about 300 million bbl, representing about 88 days of oil imports at the then-current United States import rate. Plans are to store a total of 750 million bbl in the Reserve. At the filling rate in early 1983, it was expected to be at capacity in 1990.[11]

Other countries that are heavily dependent on imported oil have also developed underground long-term crude storage facilities.

Natural gas liquids. Natural gas liquids (NGL) storage is provided at natural gas processing plants where products are delivered into the NGL gathering pipeline. Additional storage is required throughout a natural gas liquids transportation system for the same reasons crude storage is required in the crude transportation system.

Most natural gas liquids, however, must be stored in pressurized tanks or vessels. Their vapor pressure is such that they will evaporate if stored at atmospheric pressure. Typically, storage tanks for propane, butane, and similar products are horizontal, cylindrical tanks with hemispherical ends, often called *bullet tanks.*

Liquefied petroleum gases (LP-gas) are also stored in underground caverns by injecting the fluid into the cavern through wells. Withdrawal of product is also through wells extending into the cavern.

Natural gas storage. Natural gas is stored underground (Fig. 10–8), or as LNG in both aboveground and belowground tanks. Natural gas storage is needed to meet peak demands that may be much higher than the pipeline's average throughput. It would not be possible to vary production from gas wells feeding into the transmission line as widely and as frequently as demand varies. Natural gas demand is highly dependent on weather, for instance, and a method to handle these fluctuations is required.

Natural gas can be stored underground in rock or sand reservoirs that have suitable permeability and porosity. The gas is injected through wells under

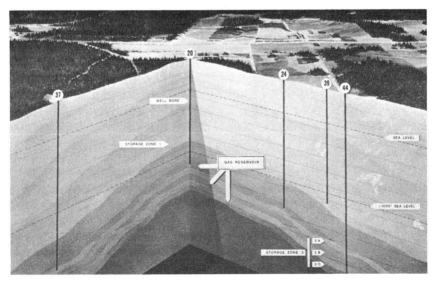

Fig. 10–8. Underground natural gas storage field. Source: *Oil & Gas Journal,* 5 September 1977, p. 115.

pressure; then pressure in the reservoir is used to force the gas out when it is needed. When demand is high, gas is withdrawn from the reservoir and is combined with gas being delivered by the transmission pipeline. Natural gas can also be stored in depleted oil or gas fields.

When demand is low, natural gas is diverted from the transmission pipeline into storage. Some of the natural gas in the reservoir must be used as "cushion" gas to allow withdrawal and injection of usable gas.

Storage reservoirs are ideally located near consuming centers and near the transmission pipeline and its compression facilities. To be suitable for gas storage, a reservoir should have the following:[12]

1. An impermeable reservoir cap rock to prevent leakage and pressure loss
2. High porosity and permeability in the reservoir rock
3. A depth sufficient to allow a safe pressure in the reservoir
4. Either no water or a controllable water condition
5. A thick, vertical formation rather than a thin, horizontal formation
6. An oil-free environment, although depleted oil-producing formations have been used

LNG.   Natural gas is also stored as a liquid. LNG storage is a way to store natural gas compactly. When liquefied at about $-260°F$, its volume is reduced to 1/600th of the gaseous volume.

LNG storage is required at *base-load plants,* complete plants that include purification, liquefaction, storage, and regasification; *terminal plants* where LNG is received from tankers and regasified as needed; and *peak-shaving plants* used to store natural gas as liquid to meet peak demands.

Underground storage tanks, aboveground storage tanks, and frozen earth storage are all used to store LNG. Because it must be stored at very low temperatures to maintain it in a liquid state, insulation is one of the most important elements of LNG storage design. In frozen earth storage, a cavity is excavated in the ground. Pipes are installed around the cavity through which refrigerant is circulated to freeze the earth and form an impermeable barrier. The cavity is topped with an insulated cover to contain the LNG. Aboveground LNG storage tanks are double walled; insulation is contained between the inner and outer walls.

Underground concrete storage tanks, also used for LNG storage, are considered applicable for large storage quantities of 1 million cu ft or more. These tanks must also be heavily insulated to prevent vaporization of the LNG while it is in storage.

Aboveground storage is used in the majority of LNG peak-shaving and base-load plants (Fig. 10–9). There are many of both types of plants around the world. Countries with no natural gas production, such as Japan, have been very

Fig. 10–9. LNG Storage tanks. Source: *Oil & Gas Journal,* 14 May 1979, p. 117.

aggressive in increasing imports of LNG in recent years to protect against high crude oil prices and crude supply interruptions.

From storage, LNG is pumped to a vaporizer that regasifies the natural gas for delivery to customers.

REFERENCES

1. Bill D. Berger and Ken E. Anderson, *Plant Operations Training: Volume 3, Gas Handling and Field Processing,* Tulsa: PennWell Publishing Co., 1980.
2. *Engineering Data Book,* 9 Ed., Gas Processors Suppliers Association, 1981.
3. Albert R. Yeats and John A. Duckett, "Metering of North Sea Oil on the Statfjord B Platform Meets Norwegian, U.K. Standards," *Oil & Gas Journal,* (18 May 1981), p. 81.
4. Press Information, Fisher Controls Inc., Marshalltown, Iowa.
5. R.E. Beaty, "API/AGA Pipeline Report: Mass Measurement Best for Ethane-Rich NGL Streams," *Oil & Gas Journal,* (15 June 1981), p. 96.
6. Wayne A. Latimer, "Mass Measurement Proves Accurate for NGL Lines," *Oil & Gas Journal,* (27 April 1981), p. 270.
7. R.H. Pfrehm, "Improved Turbine-Meter System Measures Ethylene Accurately," *Oil & Gas Journal,* (20 April 1981), p. 73.
8. *Gas Engineers Handbook,* 1 Ed., New York: Industrial Press, 1974.
9. J.W. Watson and F.A. White, "Acoustic Measurement for Gas BTU Content," *Oil & Gas Journal,* (5 April 1982), p. 217.
10. Alex Marks, *Elements of Oil-Tanker Transportation.* Tulsa: PennWell Publishing Co., 1982.

11. "SPR Filling at 220,000 b/d Toward Goal of 750 Million bbl of Crude," *Oil &
    Gas Journal,* (10 January 1983), p. 23.
12. See reference 8 above.

11

MAINTENANCE AND REPAIR

THE huge investment in pipe, pumps, compressors, drivers, and other equipment and the cost of downtime make maintenance of pipeline systems critically important.

Preventive maintenance programs for rotating and reciprocating equipment have become widespread as the result of the development of sophisticated machinery-monitoring equipment and techniques. Fast, accurate methods are available for detecting flaws in operating pipelines that could result in failure.

Combining these abilities with considerations given to reliability during design and installation, the use of effective corrosion coatings, and the selection of steel with the proper properties for a specific service has made oil and gas pipelines one of the safest industrial operations.

Operating techniques used in a modern pipeline system have also contributed to safety. With comprehensive monitoring of system conditions, leaks or other problems are quickly apparent, and their effect can be minimized.

Pipelines

A program to provide a safe, buried pipeline with the required length of service life begins in the design stage, depends on installation techniques, and requires continuous monitoring during operation. In the design stage, the proper steel must be selected to withstand temperature and pressure conditions discussed in Chapter 4 and special environments such as the Arctic and offshore. Effective corrosion coating, properly installed, is then required; if an offshore line, a well-designed concrete coating is necessary to protect coating and pipeline. Finally, the pipe must be installed using proper welding techniques and procedures to avoid damage to the coating and the pipe.

Pipeline corrosion is costly. It can result in damage to the pipeline that requires repair or replacement of pipe, loss of product through leaks, damage to property along the pipeline, and downtime.

Cathodic protection. An important approach to pipeline reliability is cathodic protection. This anticorrosion technique has long been used to protect buried pipelines from damage.

Underground corrosion of steel pipelines can result from the flow of electrical current between areas of different electric potential. This current flows from an area of higher potential through an electrolyte to an area of lower potential. The area of higher potential (the anode) will be corroded, and the area of lower potential (the cathode) will not be subject to corrosion. In the case of a buried pipeline, the soil is the electrolyte. These areas of different potential exist along a pipeline. The magnitude of the potential difference depends on soil conditions, among other factors.

In a cathodic protection system, anodes are installed and an electrical current is made to flow between the pipe and the anodes through the soil. The pipeline becomes the cathode of the system, and its corrosion is decreased. The anodes, the part of the system that is corroded, are "sacrificed."

The design of a cathodic protection system includes a current requirement survey, the selection and sizing of current drainage points, and the detailed design of the ground (anode) beds.[1]

The magnitude of the corrosion currents for a given potential difference between two electrodes (cathode and anode) depends on several factors:[2]

1. *Soil resistivity.* This is determined by temperature, moisture content, and the concentration of ionized salts present. Generally, corrosion is high in low-resistivity soils and can be low in very high-resistivity soils.

2. *Chemical constituents of the soil.* The type of salts in the soil affect the nature of corrosion.

3. *Separation between anode and cathode.* Corrosion is more likely to occur when the anode and cathode are close together. Increasing the distance between two dissimilar metals (electrodes) reduces corrosion current intensity.

4. *Anode and cathode polarization.* Protective films formed at the anode and cathode affect corrosion rate.

5. *Relative surface areas of cathode and anode.* For a given magnitude of corrosion current, the depth of corrosion on the anode will be inversely proportional to anode area.

The use of solar power for cathodic protection systems is an established commercial technology. Since power requirements are relatively low, the application fits the capabilities of solar energy systems. In addition to conserving energy, these systems can provide power in remote areas where other sources of power are difficult to install and operate. At remote locations, use of other types of power generation can require much time for servicing and maintenance, and reliability can be low. Another advantage of solar power for cathodic protection applications at remote pipeline locations is that these installations can be left unattended for long periods.

A large solar-powered cathodic protection system installed in the early 1980s in Libya serves as an example of the application of this technology.[3] The installation in Libya's Sarir field provides cathodic protection for well casings and part of a 300-mile long, 32-in. diameter pipeline from Sarir to Tobruk. In the Sarir field, pipelines connect the wells to gathering centers where the product is collected by the main pipelines. Solar power prevents galvanic action by applying direct current to the steel pipe in the wells and in the pipeline, which causes it to act as a cathode.

The Sarir system includes more than 150 anode groundbeds in the field and along the main pipeline. A typical groundbed consists of a number of silicon iron anodes; it is energized by its own multipanel solar array and a 200-amp-hr lead calcium battery bank providing a 10-day operation reserve. A cathodic protection station is self-contained, requires no external power supply, and is virtually maintenance free.

Each cathodic protection station includes solar panel arrays, electronic controls, and batteries. Panels are connected in parallel to meet output requirements; panel arrays are mounted on a steel framework facing south and at an angle of 54° to receive maximum sunlight.

Electronic controls at a typical station provide charge regulation and load regulation, and they prevent battery discharge into solar panels during darkness. The circuit consists of the regulator control circuit to sense battery voltage and a regulator switch circuit to short out the solar panels when the battery voltage reaches a critical level.

Leak detection. Oil and gas pipelines have a very good record as far as leaks are concerned. Still, interest in further improving leak detection methods and equipment is high.

Traditionally, pipelines were inspected visually by traversing the route on the ground or patrolling the pipeline route in light aircraft. Aerial inspection is still done, but emphasis now is on developing instrumentation and monitoring equipment that will provide more rapid and precise location of leaks and potential leaks. Much of this development has been spurred by an increased emphasis on pipeline safety and environmental protection.

Today's methods make it possible to detect very small leaks; the smaller the leak, the harder it is to find. The minimum size leak that can be detected depends on a number of factors:[4]

1. Type of fluid in the pipe
2. Accuracy of the metering system and accuracy of temperature and pressure transmitters
3. Line size
4. Wall thickness
5. Length of line
6. Steady-state or transient condition of the pipeline
7. Analytical equipment
8. Experience of the personnel involved

Metering accuracy plays a key role in leak detection because one important way to detect leaks is by direct observation of pressure drop and volume loss, based on comparing flow into a segment of pipeline and flow out of the segment. This approach can be effective with relatively simple instrumentation, but it is generally effective only for larger leaks. Small leaks require sophisticated instrumentation and the use of computer models of the pipeline operation.

A key factor in using a comparison of inflow and outflow to detect leaks is line fill. Line fill must remain constant for this method to be effective. Line fill, the amount of fluid to fill the line at specified conditions, is determined based on the pressure of the fluid in the line, its density, and the stretch of the pipe under pressure of the fluid. The type of product influences the type of leak detection method that can be used.

In addition to monitoring inflow and outflow to a segment of the system, other leak detection systems for liquids pipelines include acoustic emission inspection systems, instrumented pigs, and ultrasonic methods.[5] It may also be necessary to remove the liquid from the line segment and fill it with a gas to find small leaks.

Acoustic emission systems use the noise generated by gas or liquid flow from a leak in a pressurized pipeline to detect and locate leaks. This system can provide continuous leak detection by permanently installing detectors and a data transmission system; portable equipment can be used for periodic inspection. Acoustic emissions are transmitted by the pipeline and are picked up by sensors mounted at intervals on the pipe wall. Signals are carried by cable to a processor that compares the signals to those typical of normal background noise.

Instrumented pigs have also been used to monitor a pipeline for leaks. In one system, a leak-detecting pig can be moved to various positions in the pipeline by the fluid. When stopped at a test point, pressure is equalized in the test segments; if a leak exists, fluid will flow through the pig in the direction of the segment that is leaking. An electronic system transmits data through the pipe wall and soil to a surface receiver where it is analyzed by microcomputer. Other pigs for pipeline monitoring also exist.

Ultrasonic leak detection equipment includes a portable hand-held probe that is placed in contact with the bare pipe surface at prescribed intervals. Data collected are amplified by a receiver and can be analyzed audibly and on a meter.

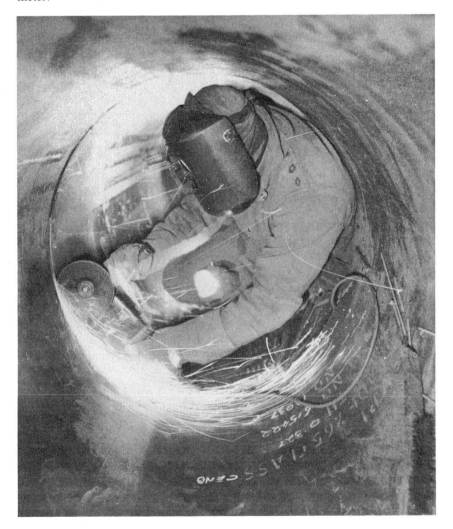

Fig. 11–1.   Welder grinds out crack to repair gas pipeline. Source: *Oil & Gas Journal,* 3 October 1977, p. 90.

Natural gas pipelines can be inspected for leaks with surface sampling instruments using the flame-ionization principle. The units consist of a sampling probe with a pump to draw an atmospheric sample to a detection cell. In the cell, the sample envelops a small hydrogen flame and carbon ions flow to a collector plate, causing an imbalance in the circuit that deflects the indicating meter. Natural gas, since it is lighter than air, tends to rise to the surface and diffuse into the atmosphere.

Gas leaks from high pressure transmission lines are relatively small. They normally form a tight, cone-shaped pattern in the soil and do not spread over a great distance. Because of this, it is necessary to place the inspection equipment accurately over the line. This type of equipment typically finds leaks resulting from stress corrosion cracks, seam leaks, and leaks in welds. When the pipe is uncovered for repair, the location of the leak may not be apparent and the equipment must again be used to pinpoint the leak. Soap testing may be used for confirmation of the leak; it is not uncommon to detect a leak that generates only a ¼-in. soap bubble in 30 sec.[5]

Leaks in natural gas liquids lines are more difficult to detect. With propane and butane, for example, it is necessary to probe in the soil and withdraw a subsoil atmospheric sample into the detector.

Pipeline repair.   When a leak is found, the method of repair varies. Some companies replace the entire joint of pipe in which the leak is found or insert a short length of pipe—a *pup joint*—where the leak exists. For some types of defects, the joint on either side of the defective joint may also be replaced. Repair of cracks in a large natural gas pipeline, for example, required unusual techniques and equipment (Figs. 11–1 and 11–2).

Fig. 11–2.   Trolley for repairs inside pipe is equipped with welder. Source: *Oil & Gas Journal,* 3 October 1977, p. 90.

Even though the system may be complex and a procedure must be carefully followed in preparing for and conducting repair operations, land pipelines are usually readily accessible. Some excavation may be required, and of course downtime can be costly if the pipeline must be taken out of service while repairs are made.

Onshore pipelines are often plugged temporarily on either side of a problem area, and flow is redirected through a bypass so work can be done on the isolated area. A variety of plugging equipment is available, and it can be applied in a wide range of situations.

In a typical plug and bypass operation, the line is uncovered and weld fittings are installed on the pipe.[6] Temporary valves are bolted onto the fittings and a hot tap is made through each, penetrating the wall of the pipe (Fig. 11–3). The tapping machine is bolted to the valve, the valve is opened, and the cutter is lowered through the valve to the pipe wall. The cutter cuts a hole in the pipe and is retrieved, along with the section removed, through the valve. The valve is then closed.

When the hot taps are complete, the temporary bypass piping can be connected to the valves. Flow is then directed through the bypass, and the damaged section of the main line is isolated. With the damaged section drained of fluid and the pressure relieved, the damaged section can be removed and the repair made.

Equipment is available (Fig. 11–4) to perform this type of operation on pipe sizes up to 48 in. diameter. Hot taps can be made into pipelines operating at pressures up to 1,400 psi and temperatures from −20°F to 700°F. Plugging can be done at pressures to 1,200 psi and temperatures from −20°F to 650°F.

In addition to mechanical plugging methods, another technique has been used in which a plug is frozen into place in the pipeline to isolate a section for repair or maintenance. When water is used to displace a hydrocarbon liquid in the pipeline, it provides a nonhazardous environment and can be quickly converted to a solid plug by subjecting the area to very low temperatures.[7] A chamber is installed on the pipeline and liquid nitrogen introduced into the chamber to cool the water below the freezing temperature. As the temperature is lowered below freezing the solid plug expands, sealing the inside of the pipe. The force resulting from expansion will be dissipated along the pipe. Also, as the pipe beneath the chamber is cooled, its diameter is reduced in the cooled area. This minimizes the effects of flaws and causes the ice plug to be forced into a conical wedge shape, increasing the effectiveness of the seal.

Leaks can also be located using the ice plug technique by isolating successive sections of the line until the leak is pinpointed.[7] Then two freeze plugs can be used to seal off the damaged section. When the repair is made, the plugs are thawed and the line is put into operation.

The repair of offshore pipelines is much more complex and repairs are much more costly. For this reason, consideration is usually given to repair methods and equipment during design of the offshore pipeline to facilitate repair when required. Formal emergency repair plans are also often made for offshore pipelines.

There is a variety of methods available for repairing submarine pipelines, but they generally fall into three categories:[8]

Fig. 11–3.   Hot tap is made in pipeline. Source: *Oil & Gas Journal*, 15 January 1979, p. 120.

1. Surface repair
2. Underwater hyperbaric welding
3. Mechanical connectors

Surface repair involves lifting the pipeline to the surface of the water and welding in a new piece of pipe to replace the damaged section, or welding flanges or fittings onto each end of the pipe after removing the damaged section. Each pipe end is then lowered back to the seafloor and is connected using a spool piece or special connector. Since the work takes place on the water's surface, the operation is sensitive to weather conditions. Also, depending on the pipe size and water depth, surface equipment required to lift the line and hold it during repair may have to be very large. Large offshore construction barges are very costly to operate.

Underwater welding techniques can be used to repair a damaged pipeline without lifting it to the surface. Welding can be done by a welder/diver while completely enclosed in a dry habitat, or the diver may work in the wet with only the work area being enclosed in a controlled environment. This method is not as sensitive to weather conditions as the surface repair method but requires skill on the part of the diver/welder.

A variety of mechanical connectors is available for underwater pipeline repair, ranging from a simple split-sleeve clamp for the repair of pinhole leaks to

Fig. 11–4.   Tapping machines come in a variety of sizes. Source: *Oil & Gas Journal*, 15 January 1979, p. 120.

complete spool pieces that can be installed without the use of divers. These connectors require some time to manufacture, and it is recommended that they be purchased and stocked so they are available when an emergency occurs.

Choice of an offshore pipeline repair method depends on location; water depth; pipeline size, age, and amount of burial; design and operating pressures; traffic in the area; special hazards such as unusual currents or mud slides; and weather conditions. The importance of the pipeline to the producing field should also be considered.

Possible alternative repair methods should be compared on a cost basis. An important point to remember in making this comparison is that the cost of the repair should be kept in perspective relative to the cost of lost production, if any. If one repair method requires minimum or no downtime but is otherwise more costly than alternative methods, it may still be easily justified if other methods require significant downtime. Even though the repair cost may be large, it often is insignificant compared with the cost of lost production.

Rotating and reciprocating machinery

Preventive maintenance is at the heart of modern pipeline system operations. Today's sophisticated monitoring and analysis equipment has significantly extended the time between overhaul or inspection of pumps, compressors, and drivers and has reduced operating costs.

Equipment is no longer operated until a failure occurs. Detailed records are kept on each equipment item in a system and its performance can be monitored continuously or checked on a prescribed schedule.

Monitoring of the performance and condition of pumps, compressors, and drivers—once depending on the experience of an operator in listening and feeling for excessive vibration or high temperature—now is done by computer-aided monitoring and analysis techniques (Fig. 11–5). Modern diagnostic methods for rotating and reciprocating equipment make it possible to determine many operating conditions that affect efficiency and fuel consumption. Potential problems can be predicted before they cause severe damage or failure.

Reciprocating equipment analysis. The engine analyzer used as a predictive maintenance tool for reciprocating engines and compressors is typical of modern monitoring equipment. In an integral engine/compressor unit, for example, an engine analyzer can indicate peak firing pressure, compression pressure of power cylinders, and suction and discharge pressure of the compressor cylinders. Vibration patterns of the working components of power cylinders and compressors can be obtained while the unit is in operation. An ignition pattern can be provided on each power cylinder to check timing and the proper functioning of the ignition system.

Fig. 11–5.   Sophisticated compressor monitoring equipment is used. Source: *Oil & Gas Journal,* 12 April 1982, p. 111.

Patterns can be displayed on an oscilloscope for study, and a permanent record can also be made of the patterns displayed (Fig. 11–6).

One program uses the analyzers for both maintenance and performance evaluation.[9] The maintenance analyzer used in this program has these basic components:

1. A crankshaft transducer that attaches to the crankshaft of an engine and gives a pulse-per-revolution signal. The signal is used to synchronize with top dead center of any crank throw on the crankshaft.
2. A signal conditioner that takes signals from a vibration pickup, pressure transducer, and ignition sensor and puts them on the trace of the pulses per revolution. Any individual signal can be put on the pulse-per-revolution trace separately.
3. An oscilloscope screen that displays the signal selected. The display screen is marked with graduations.

Everything on the display is related to the position of the crankshaft (Fig. 11–7). The 10-cm trace on the oscilloscope on this unit, for instance, represents 360° of crankshaft rotation on two-stroke engines and 720° of rotation on four-stroke engines.

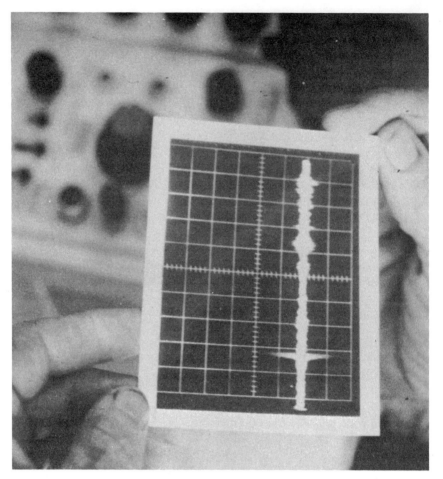

Fig. 11–6. Vibration pattern is printed. Source: *Oil & Gas Journal*, 12 April 1982, p. 111.

The performance analyzer used in this program performs all of the functions of the maintenance analyzer; in addition, it indicates pressure-volume and horsepower. The performance analyzer is equipped with a computer programmed with the necessary information to provide an indicated horsepower on a digital readout. A key use of this capability is in balancing the cylinders in a reciprocating engine. When each cylinder is producing almost the same horsepower, the fuel gas meter shows less flow to the engine.

Pressure-volume information is also useful in the compressor cylinders of a reciprocating unit. The data can be used to determine volumetric efficiency, valve loss, horsepower, and the proper functioning of valves and piston rings.

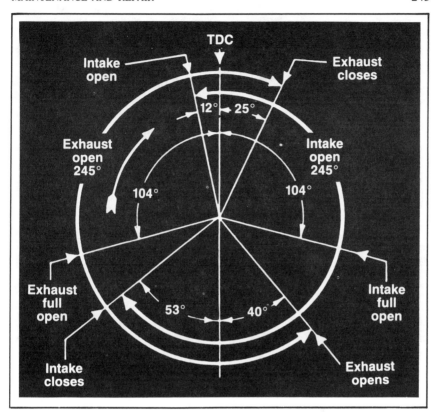

Fig. 11–7.   Instrument display is related to crankshaft position. Source: *Oil & Gas Journal,* 10 March 1980, p. 75.

A gas transmission company uses similar equipment but also uses fiberoptic probes to inspect the interior of gas turbines and sound-wave generators to detect structural flaws.[10] This company's equipment evaluation and maintenance program has drastically decreased the frequency of overhaul. Rather than completely overhauling each of its 184 reciprocating engines annually, the engine analysis and computer testing program permits some engines to run several years between overhauls.

A key element of this program is a portable computer equipped with a cathode-ray tube (CRT) display screen and keyboard. The computer can be moved to any location along the pipeline, plugged into an engine, and programmed to provide a detailed look at the engine's condition. Depending on instructions given the computer, it can check engine temperatures and pressures, measure the movement of pistons, and time the revolutions of the crankshaft. The data displayed on the CRT can be stored on a magnetic disc and converted to printed copy when a detailed analysis is required.

To evaluate the condition of gas turbines and reciprocating engines, a fiberoptic probe is also used. The instrument bends light to permit a viewer to see around corners and obstacles to the interior of a machine. The probe is inserted in turbines through small openings. The light provided by the probe is sufficient for clear visual inspection and is also adequate for television monitoring. Black and white or color pictures of a given area inside the machine can be made by attaching a camera to the outer end of the probe.

The structural integrity of bolts, rods, and other engine components can be checked with an ultrasonic detector, which is also equipped with a display screen. To test a component, a transducer is made to contact the component and ultrasonic waves generated through the component. Measurement of the speed with which the waves are reflected indicates cracks or other metallurgical flaws.

Another type of ultrasonic detector is used to check for valve leakage in cylinders. When a valve leaks, it produces a high-frequency sound that can only be detected with the use of the detector, which drops the sound to an audible range.

In this firm's maintenance program, use of the analyzers is supplemented by computer testing of lube oil and water samples. Samples are taken from each engine monthly and are analyzed to determine contamination or acidity. Samples are also checked for the presence of metal in the oil, an indication that bearings or other components are beginning to deteriorate.

Example gas-turbine maintenance program.[11]   A gas-turbine maintenance and information system using the latest analytical technqiues is part of a crude oil pipeline across Saudi Arabia. The system helps monitor and troubleshoot a large number of gas turbines and pumps along the 750-mile pipeline from a central location. The pipeline uses 60 gas turbines for pumping and power generation.

The information system is independent of the pipeline control system and allows use of special sensors to perform accurate diagnoses. The objective of the system is to help ensure the highest onstream time for gas turbines, pumps, and associated equipment.

Techniques determine rates of degradation of rotating equipment from established baselines to predict when a given gas turbine or main line pump should be taken out of service for maintenance to avoid serious problems or failure. The system (Fig. 11–8) consists of a central unit located at the pipeline control center and satellite data acquisition and processing systems at each of the 11 pump stations. The 11 remote satellite systems are connected to the central unit by microwave.

Among outputs of the system are the following:

1. System advisories, which contain important information that has just occurred and requires the operator's attention.

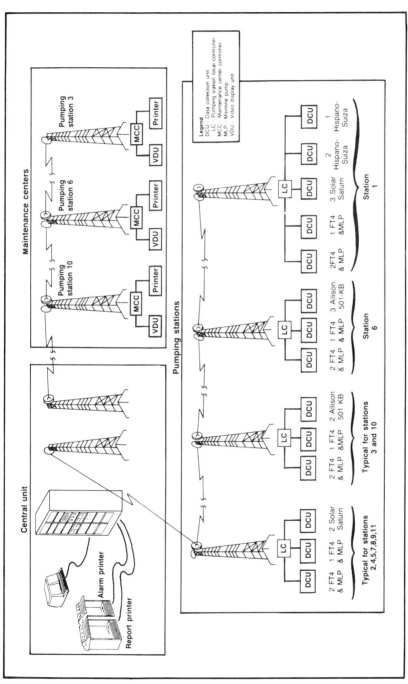

Fig. 11–8. Monitoring system configuration. Source: *Oil & Gas Journal*, 22 September 1980, p. 105.

2. Alarm advisory, printed when an alarm limit is exceeded. The message includes the present value of the parameter and the alarm limit that has been exceeded.

3. Maintenance advisories, which contain diagnostic and prognostic messages. Prognostic messages include the predicted number of remaining days in which the required maintenance action should be performed.

4. Self-health advisories, indicating problems with any information gathering unit of the system.

5. Trip report, which reviews the condition of monitored equipment prior to a station emergency-trip condition to help isolate problems.

6. Startup report, which gives the operator an indication of critical parameters to diagnose problems and provide a record of normal starts.

7. Shutdown report, generated automatically after a shutdown sequence has been completed.

8. Turbine data report, summarizing the current operating conditions and parameter values for a selected equipment item.

9. Trend history file report, which contains a file of data on the history of parameters associated with each gas turbine. Data show either the change in the parameter from an initial baseline value or the absolute value.

Reports generated by the system also include a management summary report, current alarm limits file, a serial number report, and others.

Electric motors. Electric motor drivers for pipeline pumps also require close monitoring to prevent failure. One estimate indicates that the repair cost of the "sometime" failure of an electric motor can range from 5 – 50% of the cost of a new motor.[12] Typical electric motor costs range from about $24,000 for a 500-hp motor to $47,500 for a 1,500-hp unit. In addition, motor failure can result in pipeline system downtime and the resulting cost of lost throughput. Scheduled outages to perform maintenance that prevents motor failure are therefore often recommended.

Annual documentation of pipeline pump motor operation should include a visual inspection of the motor and winding, a vibration test, and a high-voltage test. Pump efficiency testing is also recommended.

Visual inspection includes removal of end shields to expose the motor winding and other components. This inspection can require up to four hours by two men on a large motor. The vibration test is performed with the unit in operation and typically requires less than one hour unless balancing is required. A DC high-voltage test can usually be performed in less than one hour, depending on test results.

The overhaul expense involved in motor failure, plus the loss in revenue from lower throughput, can be several times the cost of the motor cleanup and revarnishing cost, for example. The cost to rewind a 1,500-hp motor has been estimated at $20,000, while the cost for a scheduled cleanup and revarnishing is estimated at about $3,000.

Other equipment

In addition to the pipeline itself and rotating and reciprocating equipment, other components of a pipeline system require constant surveillance and maintenance.

Maintenance and repair of supervisory control and monitoring systems is a critical function and requires special skills. Supervisory control systems must be reliable; the more automatic control that is provided, the more the system depends on the availability of the control system functions. The wide use of computers and sophisticated diagnostic systems, including those discussed earlier in this chapter, make expertise in maintaining these systems a critical element in efficient pipeline operations.

In pump and compressor stations, corrosion of piping and vessels must be monitored constantly to prevent failure. Changes in operating conditions may also initiate vibration problems that must be eliminated by the addition of special equipment or a change in station operating parameters. Heat exchangers used for interstage cooling and other services must be checked periodically to detect the presence and extent of fouling. Fouling of the tubes can reduce both flow efficiency and heat transfer efficiency. Periodic cleaning of heat exchangers is usually required.

REFERENCES

1. *Petroleum Transportation Handbook*, Harold Sill Bell, Editor, New York: McGraw-Hill Book Co., 1963.
2. *Gas Engineer's Handbook*. New York: Industrial Press, 1974.
3. Gordon W. Currer, "Sun Powers Libya Cathodic-Protection System," *Oil & Gas Journal*, (22 March 1982), p. 177.
4. Henry A. Brainerd and Charles W. Wilkerson, "Improving Leak Detection in Petroleum Pipelines," *Oil & Gas Journal*, (29 November 1982), p. 51.
5. Stuart B. Eynon, "Line Leak-Detection Methods Updated," *Oil & Gas Journal*, (15 September 1980), p. 205.
6. Mark B. Pickell, "Pipeline Plugging Methods Keep Pace with Industry Needs," *Oil & Gas Journal*, (3 March 1980), p. 47.
7. G.J. Howard, "Advances Made in Freezing Technique for Pipeline Plugging During Testing," *Oil & Gas Journal*, (19 April 1982), p. 108.
8. Jerry W. Woods, "Here Are Methods and Techniques for Choosing Proper Approaches to Pipeline Repair," *Oil & Gas Journal*, (19 July 1982), p. 147.

9. W.W. Graham, "Engine Analyzer Makes Key Predictive-Maintenance Tool," *Oil & Gas Journal*, (10 March 1980), p. 75.

10. James O. King and Neil Goodman, "Preventive Maintenance Keeps Compressor Engines at Peak Efficiency," *Oil & Gas Journal*, (12 April 1982), p. 111.

11. T.W. Temple and F.L. Foltz, "System Monitors Gas-Turbine Maintenance," *Oil & Gas Journal*, (22 September 1980), p. 105.

12. R.E. Wright, "Scheduled Outages to Perform Preventive Pump-Motor Maintenance Can Pay Out Fast," *Oil & Gas Journal*, (22 March 1982), p. 172.

12

TOMORROW'S TECHNOLOGY

THE search continues for better answers to pipeline construction, operation, and maintenance problems, many of which have plagued pipeline designers, builders, and operators for years. Improved technology will continue to be developed in all of these areas. Special emphasis is expected on leak detection, multiphase flow, corrosion prevention, and metering.

In addition, new challenges have been posed by the continued discovery of oil and gas in deep water and in severe onshore environments, by regulations that require special techniques to protect the environment, and by the need to improve efficiency as operating and construction costs escalate. The few areas of development highlighted here are not the only technological advances being pursued. Pipeline researchers and developers were heading in several directions in the early 1980s:

1. The use of flow improvers to increase the capacity of liquids pipelines without adding pump horsepower or looping a line.
2. More use of specialized offshore equipment, such as the reel-lay barge, to reduce construction costs.
3. For laying pipelines in very deep water, development of the vertical pipelay method.
4. Increasing use of directional drilling techniques for crossing rivers and channels.
5. Further development of underwater plows for trenching offshore pipelines to reduce their susceptibility to damage.

249

6. Techniques for more accurate measurement of both natural gas and liquids—a need brought on by the severalfold increase in petroleum product prices.
7. More sophisticated equipment and methods to monitor pipeline rotating equipment to reduce the cost of failure.
8. Increased capability of computer control systems for pipeline operation and monitoring.
9. Improved methods of pipeline leak detection to protect the environment and avoid the loss of valuable product.
10. Continued development and expanded use of automated welding techniques and equipment.

Flow improvers

Also called drag reducers, flow improvers are agents that have proven effective in increasing the capacity of crude pipelines without adding pumping horsepower or looping a line. The use of flow improvers in products pipelines is also being studied, and their application is likely to expand. Drag reducers can be particularly attractive if the need for increasing throughput capacity is not permanent.

The drag reduction phenomenon was first observed in 1945, and extensive development work was done in the 1960s and 1970s. But their cost was high at that time, relative to the cost of energy, making them uneconomical for use in pipelines because large quantities are required. Drag reducers were used commercially in the late 1970s, and development efforts aimed at improving the agents have increased.

Several compounds are being considered as drag reducers, but according to one report, the most applicable is "high-molecular-weight hydrocarbon polymers in a hydrocarbon solvent, typically 10% active ingredient by weight in a kerosine-like solution."[1]

First commercial full-scale use in large quantities was in the trans-Alaska crude pipeline, where the agent had "the consistency of cold honey." In the trans-Alaska application, several bottlenecks restricted flow. When the drag reducer was first injected at Pump Station 1, throughput increased from 1.23 to 1.28 million b/d. Injection at two other stations along the pipeline boosted flow to 1.37 million b/d. Later injection at an additional pump station increased flow to 1.52 million b/d. As an example of the application in situations in which a need for additional throughput is temporary, injection of the drag reducer at one pump station was discontinued when additional pump horsepower came onstream at that station. Use of the drag reducer in the trans-Alaska crude pipeline throughout 1980 is reported to have provided additional capacity of up to 170,000 b/d over that which would have been available without the additive.

An effective drag reducer must meet several criteria. It must be effective in small concentrations in the pipeline, able to resist degradation in transit and storage, and not have a detrimental effect on refining processes. This last criterion is important because when the drag reducer is injected into the crude pipeline, it remains in the crude as the crude enters the refinery. An important goal of further work is the development of a drag reducer that will not lose its effectiveness as it passes through pumps at pumping stations along the pipeline. Desired properties of the first commercial drag reducers deteriorated when the agent passed through pumps, and fresh material had to be injected downstream of the pumps at pumping stations.

An improved drag reducer was developed for the trans-Alaska crude pipeline in 1981.[2] The new agent is reported to be more cost effective than the first additive used, even though manufacturing cost is higher, because only 30–40% as much additive is required. According to developers, the latest version is particularly applicable where storage space is at a premium. After considerable testing in smaller-diameter pipelines, tests were run on the trans-Alaska crude pipeline with the new drag reduction agent. It was injected in amounts equal to 93, 47, 23, and 4.7 ppm (parts per million). Results indicated that for a 25% drag reduction, 36 ppm of the new drag-reducing agent was required, compared with 110 ppm of the additive used previously. Use of the new agent in the trans-Alaska crude pipeline began in early 1982, and flow rates were reported to be increased by about 200,000 b/d.

Though some equipment must be installed for injecting the agent into the pipeline, the investment in this equipment is small compared with that required to install additional pump horsepower or construct a pipeline loop.

Other applications for drag reducers have been found in pipeline operations in the North Sea and in South America. There will likely be further use of these agents as the need for more flexibility in pipeline systems increases. Supply and demand patterns change quickly, and drag reducers can offer a way to meet these changes at lower cost.

Laying methods

Use of special offshore pipelaying equipment will likely increase as the development of offshore oil and gas reserves continues to expand. The reel-lay barge will find wider application, and a new approach to pipelaying—such as the vertical lay method—will be required to install pipelines in very deep water.

The reel-lay barge concept was introduced in the early 1960s. It can provide construction cost savings because the pipe joints are welded together in a clean, comfortable environment at a shore base, are spooled onto a reel (Fig. 12–1), and then are transported to the job site and unreeled into position. Offshore construction time is significantly reduced compared with the conventional lay barge method. The equipment and technology of reel pipelaying has been

Fig. 12–1.   Pipe is spooled onto reel, unspooled at job site. Source: *Oil & Gas Journal,*
13 February 1978, p. 60.

improved, but the approach is still primarily applicable to laying smaller-
diameter lines of moderate length. There has also been some reluctance to use
the method because of fear that the flexing of the pipe resulting from spooling
and unspooling on the reel will decrease pipe strength and toughness. It has been
suggested, however, that these dangers can be compensated for by careful
selection of pipe material.[3]

Santa Fe's reel-ship *Apache* is an example of a modern reel pipelay vessel
(Fig. 12–2). It can lay pipe with diameters up to 16 in. from a reel mounted
vertically on the ship.[4] Capacity of the reel varies with the size of pipe being
spooled. About 50 miles of 4-in. diameter pipe can be spooled onto one reel;
about 5.7 miles of 16-in. diameter pipe can be wound on the reel. Rated
maximum pipelaying speed is 2 knots. This vessel also is equipped with a
saturation diving system capable of operating in water depths to 1,500 ft and a
dynamic positioning system that allows it to maintain position and move along
the pipeline route without the use of anchors.

The vertical pipelaying, or J-curve, concept has not yet been commercially
applied except in laying small-diameter flow lines from a drilling vessel using
procedures similiar to those used for handling drill string and casing during

Fig. 12–2.   Reel-lay barge installs offshore pipeline. Source: *Oil & Gas Journal,* 5 May 1980, p. 160.

drilling.[5] But a vertical pipelay vessel may be needed as future exploration in extreme water depths results in oil and gas discoveries that must be tied into processing or shipping facilities.

In conventional pipelaying from a lay barge with a stinger, the upper part of the pipeline—the overbend—is supported by the stinger. The sagbend, the lower part of the curve, must be maintained at a curvature that will not result in pipe damage by applying tension to the pipe with the lay barge tensioners. As the weight of the pipe and the water depth increase, the tension required to maintain the curvature in the sagbend below the maximum increases. Since greater water depths usually call for heavier pipe to resist external pressure, the effect of increasing water depth on lay barge tension requirements is magnified. There are practical limits to the amount of tension that can be applied to the pipe on the lay barge. Too much force applied by the tensioners can damage the pipe coating, and increasing the length of the stinger on the barge makes it more susceptible to ocean forces. The anchor system holding the lay barge in place must resist these forces.

For these reasons, the vertical-lay concept has been advanced. In the early 1980s, exploration drilling was being done in water depths exceeding 5,000 ft, making it likely that pipelines eventually will have to be installed in several thousand feet of water. Conventional pipelaying techniques with minor modifications have been used to lay pipelines in water depths as great as about 2,000 ft, but the forces involved at these depths appear to be near the limits of conventional horizontal-lay methods.

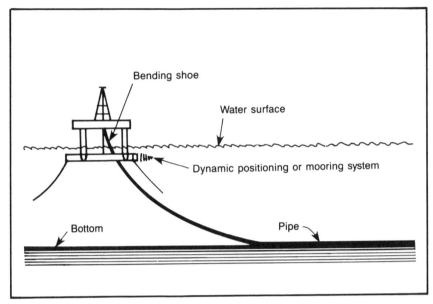

Fig. 12–3.   Vertical pipelay concept.

With a vertical-lay vessel, pipe would be suspended vertically (Fig. 12–3). A new joint would be added, aligned with a line-up clamp, and welded in the vertical position. The pipe would then be lowered so the next joint could be added. Weld inspection and coating stations could be included at different levels on the vessel. The pipeline would exit vertically through a *moonpool*, a vertical opening in the vessel hull common in offshore drilling vessels.

Not only would total tension required on the pipe during laying be less than when laying from a horizontal position, but the bulk of the tension force would be in the vertical direction and would not have to be resisted by the vessel anchors. In fact, the small magnitude of this force might make the use of dynamic vessel positioning practical for vertical-lay vessels. Forces on equipment suspending the pipeline during laying would be well within the capacities of derricks used on offshore drilling rigs. For instance, calculations show that to lay a 36-in. pipeline in 3,000-ft water depths, about 350,000 lb of tension must be applied in the vertical-lay method. By comparison, laying the same size line in the same water depth using a horizontal-laying approach would require more than 1 million lb of tension.[5]

Whether or not this approach to deep water pipelaying becomes commercial depends on the economics of a particular project. Nevertheless, it does offer a potential alternative in extreme water depths.

In 1982, Total-CFP, a French firm, reported its J-curve pipelaying method had passed all land tests and was scheduled to be used on a drillship to install an

1,800-meter-long underwater line in the Mediterranean in 1983.[6] That system is designed to lay pipe in water depths to 3,000 meters using a dynamically positioned drillship or vessel. An electron beam welding chamber is used in the system, which can weld 24-in. diameter pipe with a 1½-in. wall thickness in about 3 minutes.

Directionally drilled crossings

Horizontal directional drilling has been used for several years to cross waterways with pipelines. The technique is well proven, having been used in more than 100 crossings by the early 1980s. But the method is expected to see even greater use.

It offers a number of advantages over dredging or other stream-crossing techniques used in the past. When crossing waterways in which traffic is heavy, there is no disruption of traffic; environmental damage is minimal since neither the stream bed or its banks are disturbed; permits are easy to obtain because there is little environmental impact; and the pipe can be placed well below any future channel deepening, avoiding the need to be relocated.

An example of the technique (Fig. 12–4) providing many of these advantages was a crossing of the Houston Ship Channel, where nearly all the pipeline crossings since 1976 have been installed by horizontal directional drilling.[7] The first crossing of the channel was with 1,400 ft of 8⅝-in. diameter pipe in a 10¾-in. diameter casing. The low point in this crossing is 95 ft below mean low tide, deeper than most other lines in this area. Channel depth at that point is 40 ft below mean low tide. Regulations require a pipeline to be at least

Fig. 12–4. Horizontal directional drilling for pipeline. Source: *Oil & Gas Journal,* 24 March 1980, p. 161.

17 ft below the authorized channel depth, and this depth must extend at least 25 ft beyond the edge of the channel.

Another job in Venezuela involved a 4,550-ft crossing under the Orinoco River with 22-in. diameter pipe. Other horizontal drilling crossings have been completed by the same firm in the United States, Mexico, the Ivory Coast, and the United Kingdom.[8]

The key to the method is a steering tool that guides a drilling bit along a prescribed path under the stream or channel. The steering system consists of an instrument package placed in the drillstring behind the downhole drilling motor. The instruments measure direction, angular inclination, and rotation of the tool face. The drilling motor is rotated by pumping drilling fluid through the motor, which in turn rotates the drill bit.

Reducing the drilling fluid pressure activates the instruments, which then transmit data from the bottom of the hole to the surface by low-frequency waves. At the surface, the data are fed into a computer that calculates and displays the location of the downhole tool. The actual position of the bit can be compared to the desired path of the hole and changes can be made in direction if necessary.

The surface drilling rig is placed on one side of the stream or channel and the hole is begun at an angle to the horizontal. In the Orinoco crossing, the entry of the hole was at about 12° above the horizontal. Rather than beginning from the vertical position, as is the case when drilling an oil or gas well, the directional pipeline crossing unit is nearly horizontal. As the bit drills the length of a pipe section, another section of pipe is added at the surface, and drilling proceeds.

In the Orinoco crossing, a pilot hole was first drilled. Then the hole was reamed by making two passes with larger bits until the required hole diameter was obtained. The exit point on this job was at the projected distance and less than 12 meters from the target.

The ability to monitor the position of the bit almost continuously, relative to the desired position, and to make changes to stay on course is the key to the effectiveness of this technique. Advances in position monitoring and direction control are sure to make this pipeline crossing method even more practical and economical in coming years.

Underwater plows

The development of underwater plows for burying offshore pipelines below the seafloor has advanced rapidly since the mid-1970s. The technique can often provide more effective burial more economically than other burial methods, especially for long offshore pipelines. It is less costly than jetting, for example, in many cases. Modern plowing techniques can also provide better burial in a variety of soil conditions, making the pipeline more stable and reducing the possibility of damage during its operating life.

Most commercial plows built to date have been designed for a specific project, in part because a detailed knowledge of the pipeline route is necessary to design the plow shares and other components properly. Continued development, however, is resulting in plows that can trench in a variety of soil conditions.

The general types of plows used to date include the pretrenching plow, the post-trenching plow, and the simultaneous plow. The pretrenching plow was the first type to be used. The first commercial plow of this type for use on a large-diameter pipeline was developed in 1976–1977. The pretrenching plow is towed along the pipeline route before the pipeline is laid and cuts a trench for the pipeline to lie in. Post-trenching plows are used after the pipeline has been laid on the ocean floor. The plow follows the pipeline and cuts a trench beneath the pipeline. The pipeline then settles into the ditch behind the plow. A simultaneous plow has also been developed that is used in conjunction with the lay barge and allows pipe laying and trenching at the same time.

The split-share plow was developed in the late 1970s. This design allowed a plow to be of lighter construction. In an early commercial design of the split-share plow, an instrument panel mounted on the plow was monitored by an underwater TV camera. Data supplied by the instruments included plow load, trench depth, plow share position, and the angle that the pull cable made with the pipeline.

It is likely that development of more effective plows and equipment that can handle a wider variety of conditions will continue. Recent advances have focused on designs that can trench in both soft and hard soils, promising more versatile equipment in the future. Other developments in this area include a plow that can trench effectively in uneven sea beds (Fig. 12–5).

Measurement

Traditional measurement equipment—orifice meters and turbine meters—will continue to be the heart of oil and gas measurement systems for the foreseeable future. But other types of meters are in commercial use and are expected to find increasing application.

There will also be increased emphasis on the measurement of natural gas heating value. Mass measurement rather than volumetric measurement will be used increasingly, especially to measure light hydrocarbons and other streams whose density varies widely with flowing conditions.

Important improvements in oil and gas measurement will also focus on the use of sophisticated equipment to improve the accuracy of conventional measuring devices and drastically increase the speed of data collection and analysis. Today, the value of oil, natural gas, and products is such that sizable expenditures to reduce inaccuracy and loss can be justified.

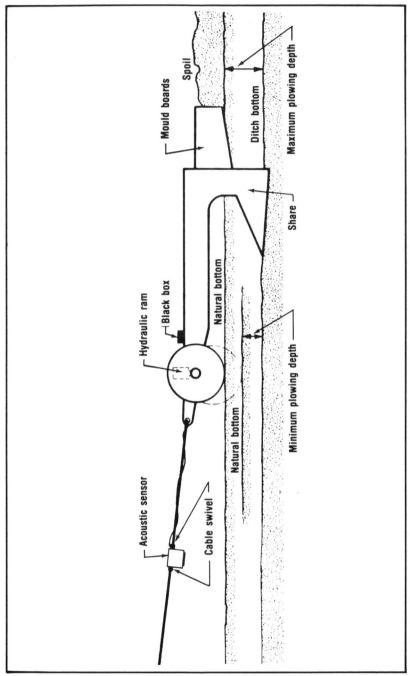

Fig. 12–5.   Underwater plow for uneven seabed.   Source: *Oil & Gas Journal*, 4 May 1981, p. 133.

The use of microcomputer-based data gathering and analysis in measurement systems is already showing good results. An automatic metering system, including multiple meter runs and a meter prover operated by a small computer, has proved successful in an offshore tank loading application. This system automatically conducts up to 10 proving runs on a specified meter and calculates the average meter factor. The computer warns the operator if flowing conditions vary beyond preset limits, and the operator can initiate a meter proving operation.

In mass measurement, use of a computer permits virtually instantaneous integration of data from a turbine meter and a density meter to provide flow quantity.

New types of meters, such as the vortex flow meter discussed in Chapter 10, are expected to be used in a growing number of applications. They can operate over a broader range of flows than the orifice plate/differential pressure meter with acceptable accuracy, and they require no density correction in determining flow volume.

Monitoring and control

Wide use is already being made of computer-based monitoring and control of all types of pipeline systems, but this area of oil and gas pipeline operations is due for more development. Advances will be fueled by the increasing capability of computer hardware and growing pressures to reduce operating and maintenance costs.

No longer will such a control system be used only on large, complex systems because it is the only feasible way to maintain control and gather information. More and more, computer-based operation can be justified for even small field gathering systems. Real-time data and the ability to make repetitive calculations easily can provide fuel savings and reduce labor costs in even small pipelines. Though the change in operating conditions may be slight, the accumulated savings over a year's operating time can be significant.

Coupled with system control, real-time monitoring of rotating equipment and other system components can reduce fuel consumption. Sophisticated evaluation of pump, compressor, and driver mechanical condition has been shown to reduce maintenance costs and extend operating life.

Computer software will also become more application-oriented and practical. Pipeline operators and designers now use computer equipment routinely. As software capability grows, they can find ways to apply the latest methods quickly. Many modern pipeline systems include scores of inlet points and delivery points as well as several different products. Only a computer-based system of control, measurement, and equipment monitoring can hope to operate such a system efficiently.

Leak detection will also benefit from advances in computer modeling of pipeline systems. Some methods of leak detection require precise knowledge at all times of a variety of parameters. The only way to obtain these on a timely basis is with continual scanning by instruments that report to a computer. The computer can then calculate the volume of fluid entering a segment of the pipeline and compare it to the volume leaving the segment and determine if a leak has occurred. There will be increased use of pipeline system modeling for leak detection and to improve efficiency. Modeling will also be common in solving complex design problems, such as those involved in two-phase flow.

The result of this widespread use of computer-based monitoring, control, and modeling will be greater efficiency and lower operating costs.

Welding

Though it is still used for only a small portion of oil and gas pipeline welding, automated welding has provided pipeline builders and operators with high-quality welds and high pipeline laying rates during construction. But continued development of automated welding equipment and techniques will be needed to meet the demands of tomorrow's construction environments.

A computer-controlled welding machine designed for multipass welding of heavy-wall pipe was developed in 1974, for instance, and underwent testing and development in the following years. This machine was developed for lay barge operation, then was tested to evaluate its effectiveness for use in land pipelaying. In offshore tests, it was operated at system rates equal to more than 200 single joints per day for 36-in. pipe with a 0.750-in. wall thickness.[9]

The computer-controlled machine has four welding heads, and the computer is programmed for all weld passes needed to make a complete weld at a single station. It can also be used with a multistation welding arrangement where any one station can make any weld pass. Any welding parameter can be changed at any position by programming the computer and the appropriate transducer.

This computer-controlled system uses a hot-wire, gas-tungsten arc welding (GTAW) process, and all welding is done from the outside of the pipe. An internal line-up clamp is used to ensure the proper positioning of the two pipe joints. Each of the four welding heads in the unit has an electronic controller. A central processing unit controls the interaction of the heads. Controls for one welding head include an automatic voltage control chassis, wire feed and AC heat control chassis, DC current control chassis, and an oscillator chassis.

All weld parameters that vary either from pass to pass or during a pass are stored in magnetic core memory. DC arc voltage, DC current, wire speed, wire heat, oscillator rate, oscillator width, and carriage rate are analog set points to their respective control circuit.

This particular system underwent further development in the late 1970s. It shows the potential to increase the efficiency and quality of pipeline welds even under difficult field conditions.

Special welding processes will also be needed for possible future pipeline operations. In the J-curve, or vertical-laying, method designed for installing pipelines in a very deep water, the electron beam welding method is thought to be the best approach. In the J-curve method, welding is expected to be done at a single station, and electron beam welding offers an advantage in this arrangement. It is faster than other welding processes and can weld thick-wall pipe in a single pass. No preheating or postheating is required when using electron-beam welding.

The future

Pipelines will be the most economical way to transport oil, gas, and petroleum products overland in the foreseeable future. It is apparent from the previous chapters that the oil and gas industry has the basic technology needed to build and operate pipelines in almost any environment. Emerging technology discussed in this chapter involves only refinements of the capabilities already routinely used in oil and gas pipeline operations.

Most future advances in oil and gas pipeline technology will also be gradual improvements in efficiency and capability. Few dramatic changes are expected.

Welding will continue to be the best method for joining pipe, for instance, although new procedures will be developed. Offshore pipelines will continue to be installed from lay barges, but lay barge capability will be increased to handle deeper water.

The continuing challenge facing pipeline designers and operators is to reduce construction costs and improve operating efficiency. Most of the technology required in the coming years will not be revolutionary, but the need for innovative approaches to lower cost and higher efficiency has never been greater.

REFERENCES

1. Jack B. Holt, "Drag Reducers Boost Crude-Line Throughput," *Oil & Gas Journal*, (19 October 1981), p. 272.
2. W.R. Beaty, et al., "New High Performance Flow Improver Offers Alternatives to Pipeliners," *Oil & Gas Journal*, (9 August 1982), p. 96.
3. Parviz Mehdizadeh and E.Y. Chen, "How Reeled-Pipe Lay Method Affects Pipe," *Oil & Gas Journal*, (4 February 1980), p. 64.
4. Svend Jorgensen, "Flowlines Laid by Reel-Ship *Apache*," *Oil & Gas Journal*, (5 May 1980), p. 160.

5. W.J. Timmermans, "Vertical-Lay Vessel May Cut Offshore Pipeline Cost," *Oil & Gas Journal,* (9 July 1979), p. 144.
6. "Laying Large Diameter Pipe in 3,000 Ft Water," *Ocean Industry,* (December 1982), p. 58.
7. Hugh W. O'Donnell, "Directional Drilling Eases Ship Channel Pipeline Crossings," *Oil & Gas Journal,* (24 March 1980), p. 161.
8. "Pipeline Record Claimed in Orinoco Crossing," *Oil & Gas Journal,* (19 April 1982), p. 54.
9. G.J. Merrick and G.E. Cook, "Computer-Controlled Offshore Welding System Being Tested," *Oil & Gas Journal,* (16 January 1978), p. 78.

Index